For Grade 4

Math on Target

Parent/Teacher Edition

Using Thinking Maps to Solve Multiple-Choice, Short-Answer, and Extended-Response Problems

Written By:
Yolande F. Grizinski, Ed.D.
Leslie Holzhauser-Peters, MS, CCC-SP
Claire L. Crook, Ph.D.

Show What You Know® Publishing

Published By:
Show What You Know® Publishing
A Division of Englefield & Associates, Inc.
P.O. Box 341348
Columbus, OH 43234-1348
1-877-PASSING (727-7464)

www.showwhatyouknowpublishing.com

Copyright © 2006 by Englefield & Associates, Inc.

All rights reserved. No part of this book, including interior design, cover design, and icons, may be reproduced or transmitted in any form, by any means (electronic, photocopying, recording, or otherwise), without the prior written permission of the publisher. Permission to reproduce pages 123–137 is granted by the publisher to one teacher for use with students in a single classroom.

Printed in the United States of America
08 07 06   20 19 18 17 16 15 14 13 12 11 10 9 8 7 6 5 4 3 2 1

ISBN: 1-59230-160-6

Limit of Liability/Disclaimer of Warranty: The authors and publishers have used their best efforts in preparing this book. Englefield & Associates, Inc., and the authors make no representation or warranties with respect to the contents of this book and specifically disclaim any implied warranties and shall in no event be liable for any loss of any kind including but not limited to special, incidental, consequential, or other damages.

About the Authors

Yolande F. Grizinski received a Bachelor's degree from Miami University, a Master's degree from Wright State University, and a Doctorate in Education from the University of Cincinnati. She has worked in public education for 30 years as a curriculum consultant in the areas of language arts with a focus on writing assessment. She is currently the Assistant Superintendent of the Warren County Educational Service Center in Lebanon, Ohio.

Leslie Holzhauser-Peters holds a Bachelor's degree from the University of Cincinnati and a Master's degree from Miami University. She has 27 years of experience working in public schools in Special Education as a Speech-language pathologist, as a Supervisor, and currently as a Curriculum Consultant. Her areas of expertise are language, literacy, and intervention.

Claire L. Crook holds a Bachelor's degree from The Ohio State University, a Master's degree from Xavier University, and a Doctorate in Mathematics Education from The Ohio State University. She has more than 30 years experience in public education as a teacher, administrator, and consultant. She is currently a Mathematics Curriculum Consultant.

The authors met at the Warren County Educational Service Center in Lebanon, Ohio. There they developed and implemented a host of language arts and mathematics initiatives.

Acknowledgements

Show What You Know® Publishing acknowledges the following for their efforts in making this assessment preparation material available for students, parents, and teachers.

Cindi Englefield, President/Publisher
Eloise Boehm-Sasala, Vice President/Managing Editor
Lainie Burke Rosenthal, Project Editor/Graphic Designer
Erin McDonald, Project Editor
Christine Filippetti, Project Editor
Jill Borish, Project Editor
Charles V. Jackson, Project Editor
Heather Holliday, Project Editor
Jennifer Harney, Illustrator/Cover Designer

Table of Contents

Foreword viii

Introduction 1
   The Purpose of *Math on Target* 1
   Ways to Use *Math on Target* 2
   Literature Connections 2
   What Is a Mathematics Thinking Map? 3
   Guidelines for Using the Thinking Map 4
   The Five Mathematics Standards 5

Preface One: Using Thinking Maps—The Model Lesson 7
   The Model Lesson 7
   Defining Each Component of the Model Lesson 7
   Evaluation of Student Work 11
   Use of Rubrics to Assess Student Work 12
   Model Lesson: Putting it All Together, Step-by-Step 14

Preface Two: Design Your Own Higher-Level Mathematics Problem 16

Chapter 1: Number Sense 17
   The purposes of this chapter 17
   Teaching tools included for extending students' thinking with
      Number Sense 17
   What is Number Sense? 17
   What does Number Sense look like? 18
   Vocabulary related to Number Sense 18
   Teaching Tips 19
   Model Problem 1: Multiple-Choice 20
   Practice Problems 1–7 22
   Model Problem 2: Short-Answer 29
   Practice Problems 8–9 31
   Model Problem 3: Extended-Response 33
   Practice Problem 10 35
   Literature Connections for Number Sense 36

© Englefield & Associates, Inc.   COPYING IS PROHIBITED   v

Table of Contents

Chapter 2: Measurement 37
    The purposes of this chapter 37
    Teaching tools included for extending students' thinking with Measurement ... 37
    What is Measurement? 37
    What does Measurement look like? 38
    Vocabulary related to Measurement 38
    Teaching Tips 41
    Model Problem 4: Multiple-Choice 42
    Practice Problems 11–17 44
    Model Problem 5: Short-Answer 51
    Practice Problems 18–19 54
    Model Problem 6: Extended-Response 55
    Practice Problem 20 57
    Literature Connections for Measurement 58

Chapter 3: Geometry 61
    The purposes of this chapter 61
    Teaching tools included for extending students' thinking with Geometry 61
    What is Geometry? 61
    What does Geometry look like? 62
    Vocabulary related to Geometry 62
    Teaching Tips 64
    Model Problem 7: Multiple-Choice 65
    Practice Problems 21–27 67
    Model Problem 8: Short-Answer 74
    Practice Problems 28–29 76
    Model Problem 9: Extended-Response 78
    Practice Problem 30 80
    Literature Connections for Geometry 81

Table of Contents

Chapter 4: Algebra 83
    The purposes of this chapter 83
    Teaching tools included for extending students' thinking with Algebra 83
    What is Algebra? 83
    What does Algebra look like? 84
    Vocabulary related to Algebra 84
    Teaching Tips 85
    Model Problem 10: Multiple-Choice 86
    Practice Problems 31–37 88
    Model Problem 11: Short-Answer 95
    Practice Problems 38–39 97
    Model Problem 12: Extended-Response 99
    Practice Problem 40 101
    Literature Connections for Algebra 102

Chapter 5: Data and Probability 103
    The purposes of this chapter 103
    Teaching tools included for extending students' thinking with
        Data and Probability 103
    What is Data and Probability? 103
    What does Data and Probability look like? 104
    Vocabulary related to Data and Probability 104
    Teaching Tips 105
    Model Problem 13: Multiple-Choice 106
    Practice Problems 41–47 108
    Model Problem 14: Short-Answer 115
    Practice Problems 48–49 117
    Model Problem 15: Extended-Response 119
    Practice Problem 50 121
    Literature Connections for Data and Probability 122

Chapter 6: Manipulatives 123

Foreword

We are fortunate enough to have worked with many wonderful educators and students in the years before writing this book. Each of us has observed mathematics classrooms and students involved in solving mathematics problems. With the inception of standards-based assessments and high-stakes testing throughout the nation, students are asked to show what they know through written responses to mathematics problems.

In our work, we have observed that many students attempt to solve mathematics problems of various formats without having a systematic approach to use. To improve students' problem-solving processes, we have developed a thinking map that students can use to solve most word problems in a systematic way.

Five mathematics strands tied to national standards are included in this book—Number Sense, Measurement, Geometry, Algebra, and Data and Probability. The purpose of this book is to provide parents, teachers, and students with model problems for multiple-choice, short-answer, and extended-response problem types. For each of the five mathematics chapters, there is one multiple-choice model with seven practice problems, one short-answer model with two practice problems, and one extended-response model with one practice problem. As a result, teachers have access to more than 60 problems to model the thinking that students need to solve word problems.

This book was written for students who are working at the fourth-grade level.

A model lesson framework outlines the steps for preparing students for a lesson using the problems included in *Math on Target*. The steps include determining prior knowledge and connecting to previously learned concepts, modeling the thinking map and problem-solving strategies, guiding practice by observing students, and practicing independently.

Included in both the Parent/Teacher Edition and the Student Workbook are blackline masters of manipulatives that include pictures of two-dimensional figures, one-inch and half-inch grids for problem solving, measuring tools (a protractor and inch and centimeter rulers), spinners, cube models to show volume, and a clock face. These tools as well as others that you use in your classroom can help students understand abstract mathematical concepts.

At the end of each chapter, you will find a list of suggested books that could be used to illustrate mathematical concepts or to create interdisciplinary experiences.

Introduction

The Purpose of *Math on Target*

By the time students reach fourth grade, they are expected to analyze and solve multi-step problems. Students are expected to explain the strategies they use and to describe change. The ultimate goal is to develop each student's ability to apply the mathematics learned in school to everyday, real-world situations. *Math on Target* was developed for the following purposes:

1. To provide students with a systematic way to look at word problems. By using the thinking maps provided, students will develop the ability to explain their problem-solving strategies and take their thinking to a deeper level.

2. To provide teachers with problems linked to national standards in three assessment formats—multiple-choice, short-answer, and extended-response. These standards include
   - Number Sense
   - Measurement
   - Geometry
   - Algebra
   - Data and Probability

In addition to the two major purposes of *Math on Target*, the thinking map can help teachers in many ways, including the following:

- Analyzing where students break down in their knowledge of facts, skills, and mathematical processes that have already been taught in the regular mathematics program;
- Providing sample problems for modeling, practice, and diagnostic assessment purposes;
- Presenting problems that help students to develop the conceptual understanding and higher-level thinking skills required of national standards: problem solving, reasoning and proof, communication, connections, and representations; and
- Providing a format for students to show their thinking and work as they transfer their thoughts into a written format for a reader.

Introduction

Ways to Use Math on Target

Use *Math on Target* as

1. A major component of a math response journal program. Students would complete multiple-choice, short-answer, and extended-response problems in the five mathematics standards areas.

2. A pre- or post-assessment for different types of problems.

3. A diagnostic tool to identify students' progress in their problem-solving skills.

4. A bank of problems that support classroom instruction.

5. Support for concentrated standardized test preparation programs.

6. Part of a summer school math program or intervention program.

7. A benchmark assessment to determine individual student achievement.

8. Math problems for individual student practice.

Literature Connections

At the end of each chapter, you will find a list of suggested books that you may want to use as part of a model lesson. During the model lesson, as you are building on prior knowledge and making connections, you may want to use the books or sections of the books listed at the end of each chapter. These titles may be used to make interdisciplinary connections as you develop lesson plans.

The books are suggested to illustrate mathematical concepts and are at varied reading levels. Some titles may need to be read to the students, while others may be read and explored by individual students or cooperative student groups.

What Is a Mathematics Thinking Map?

The thinking maps in *Math on Target* provide students with a systematic way to think through any mathematics problem. The use of a thinking map prompts students to reread and rethink what it is they need to solve. The thinking map helps students to

- Review the information and pick out the elements that are important (such as vocabulary and what the problem is asking).
- Reread the problem critically several times.
- Determine what problem-solving strategies are needed in either mental or written processes.
- Develop the ability to explain their answers.
- Check to see if an answer makes sense as a final step of the problem-solving process.

The thinking map can be used to solve any mathematical word problem that grade four students encounter.

One of the greatest challenges students face is to read and reread problems critically. The thinking map enables students to trace their problem-solving strategies so they can describe their thinking and ultimately demonstrate their understanding.

The problems that are included in *Math on Target* have these qualities and characteristics:

- Real-life situations
- Various methods for arriving at a correct solution
- Multiple steps
- Higher-level thinking
- Skills, concepts, and processes that are part of the regular math program
- Problem-solving processes that can be traced and explained with symbols, pictures, and words

Introduction

Guidelines for Using the Thinking Map

1. Model the thinking map by working through the problem-solving process of a sample problem.

2. Explain the purpose of the thinking map to the students.

3. Model the completion of the thinking map:
   - Read the problem.
   - Reread the problem.
   - Find the important vocabulary.
   - Reread the problem.
   - Write a sentence or phrase that clarifies what the problem is asking you to solve.
   - Reread the problem.
   - Orally discuss the problem, act it out, and/or use manipulatives.
   - Determine whether you want to use a picture, a calculation, an equation, and/or a formula.
   - Use words, numbers, and/or pictures to explain the answer.
   - Determine whether the answer makes sense.

   > Show the students that they need to read and reread the problem many times. A good mathematician's work is not based on speed, but on accuracy.

4. Discuss that there is more than one way to solve the problem.

5. Share the scoring rubrics with the students.

6. When presented with an assessment question, model the following for the students:
   - How they can mentally use the thinking map, and
   - How they can use words, pictures, and/or numbers to explain their mathematical thinking to a reader.

   > The ultimate purpose of *Math on Target* is to teach students to internalize a thinking map that will guide them through the thinking process for any mathematics problem.

The Five Mathematics Standards

Standard	Standard Description	Problem Topics
Number Sense	ratio concept of "one" equivalent forms rounding to the nearest half numbers less than zero order of operations common denominators place value estimation strategies	part-to-part, part-to-whole (percent, part-to-whole) equivalence of fractions fractions, decimals, percents decimals, fractions, mixed numbers use number lines and applications (e.g., temperature) commutative, associative, distributive, and inverse properties, use of parentheses add and subtract fractions add and subtract decimals whole numbers, fractions, decimals, flexible rounding, and front-end strategies
Measurement	select appropriate tools to measure angles U.S. customary units and metric measurement units develop formulas differences among units 3-dimensional objects compare lengths	use tools to measure and draw angles conversions within the same measurement system perimeter and area of triangles, rectangles, and parallelograms, and volume of rectangular prisms linear, square, and cubic units surface area and volume points on a grid, coordinate plane

Note: Mathematical processes are embedded throughout all five mathematical standards.

Introduction

The Five Mathematics Standards

Standard	Standard Description	Problem Topics
Geometry	circles standard language parts of angles congruence physical models coordinate system measurements of angles polygons	radius, diameter, center, circumference line, segment, ray, angle, skew, perpendicular vertex, rays, interior, exterior use properties to solve problems sum of interior angles of triangles and quadrilaterals values of negative numbers (x or y) degrees of rotation
Algebra	use calculators and computers to develop rules for patterns and functions describe patterns and relationships represent problems draw conclusions and make predictions understand quantitative changes	use physical materials, representations, words, tables, graphs use variables as unknown quantities equations and inequalities use model problems, physical materials, and visual representations, including models, graphs, and tables how changes in variables affect the values of other variables
Data and Probability	construct and interpret match appropriate graph to data sets read and interpret complex data displays collect appropriate data; modify and communicate findings as appropriate measure central tendencies determine probability of an event make and justify predictions	frequency tables, circle and line graphs numerical vs. categorical discrete vs. continuous double bar graphs pose questions, collect and display data, communicate findings, and justify interpretations range, mean, median, mode ratios, fractional notations simple experiments

Note: Mathematical processes are embedded throughout all five mathematical standards.

Preface One

Using Thinking Maps—The Model Lesson

The Model Lesson

There are four major components that are key to any good lesson that incorporates a mathematical thinking map. These four components include:

1. **Prior knowledge**—to determine the students' prior knowledge about the standard, content of the problem, and/or mathematical process of the problem.

2. **Model**—to model the use of the mathematical thinking map through class discussion, with the thinking process modeled aloud as the thinking map is completed.

3. **Guided practice**—to provide guided practice by presenting a problem for students to complete that matches the standard, content, or process you want to teach.

4. **Independent practice**—to provide independent practice so students are provided with multiple-choice, short-answer, or extended-response problems.

Defining Each Component of the Model Lesson

I. **Determining Prior Knowledge of Students**

Students must make personal connections with the mathematics problems being presented. Establishing a student's familiarity with the material is important for interpreting and comprehending meaning and ultimately retaining information. The following questions can be used with any mathematics lesson to tap into the prior knowledge of the whole class or an individual student:

1. What do you know about _____? (The student's response provides the teacher with an assessment of what the student knows prior to the lesson.)

2. Where do you see this (or use this) in your life? (The student's response helps to establish the importance of the problem for now and for the future.)

Both of these questions show how students are connected with the material. This process is important for the retention of information. Students need to make connections themselves.

Preface One—Using Thinking Maps—The Model Lesson

> As an example, these sample questions could be used to tap prior knowledge of the Measurement Standard:
> - Does anyone know what measurement is?
> - What do we know about measurement?
> - What are some measurement units that you use every day?
> - Who uses measurement?
> - Where do you see measurement problems every day?
> - What does measurement mean?
> - What are measurement tools?

II. Modeling

Modeling the use of the mathematical thinking map is an important component of any lesson. Teacher questions that can be used while modeling each section of the thinking map are shown on the next page.

The teacher must demonstrate the thinking process for the use of the mathematical thinking map frequently because it is the part of the lesson when students construct and extend meaning.

There are three types of modeling:

1. Demonstrating the thinking process used in completing the mathematical thinking map

2. Examining the different strategies that students have used to solve the same problem

3. Providing samples of acceptable, incomplete, and unacceptable student responses; correlations should be made between samples and a scoring rubric (The scoring rubrics can be found on pages 12–13.)

All three types of modeling must be done frequently. Time is well spent during the modeling phase of the lesson. Modeling clarifies the task and enables the student to internalize the standard.

Thinking Map with Teacher Questions

Read the Problem	☐ Read the Problem
Reread the Problem	☐ Reread the Problem
Write the important math vocabulary that tells you what to do.	What words do you understand? Do you see any words or symbols you don't understand? What words provide you with the information to know what to do?
Reread the Problem	☐ Reread the Problem
What information do you have that you can use to solve the problem? Can you get clues from: ☐ The answer choices ☐ Pictures, charts, or graphs ☐ A problem you have solved before	What are the important numbers or symbols? Does the problem give you more information than you need?
Reread the Problem	☐ Reread the Problem
Solve the problem. Use one or more: ☐ Act it out. ☐ Use manipulatives. ☐ Do a calculation: addition, subtraction, multiplication, or division. ☐ Draw a picture, graph, or table. ☐ Set up an equation. ☐ Write a formula.	Why did you decide to use this method? Can you think of another method that might work? Is there a more efficient strategy you could use? Do you see a pattern?
Use words, pictures, or numbers to explain your answer.	Does anyone have the same answer but a different way to explain it?
Does your answer make sense? Why or why not?	Do you think your answer would work with different numbers? Can you convince the rest of the class that your answer makes sense?
Answer the Problem	☐ Be sure to record your answer on the previous page

Preface One—Using Thinking Maps—The Model Lesson

> These generic questions can be used at any time in the modeling of the thinking map:
> - What do you think about what _____ said?
> - Does that make sense?
> - Can you explain…?
> - What do you notice?

The model problem needs to match the purpose of the standard, content, or process. Models of word problems can be created using information from a variety of sources:

- Science experiments
- Art projects
- Newspapers
- Recipe books
- Textbooks
- Student-generated problems

Over time, develop a bank of student responses that are good examples of the different ways students have solved the problems of *Math on Target*. Include different levels of student responses based on the scoring rubrics.

III. Guided Practice

During guided practice, the teacher should provide various levels of support to students as needed.

Provide a problem for students to solve that matches the skill you are teaching in the most real way possible.

Use problems from *Math on Target*.

After considerable guided practice, provide students with many opportunities to engage in independent practice with real-world problems.

IV. Independent Practice

Independent practice is an opportunity for students to complete problems on their own. During independent practice, students should be given a thinking map, a problem from *Math on Target*, and paper. It is important for teachers to provide enough time for students to complete each problem.

During independent practice, no teacher or peer assistance should be provided for students. It is critical that students be provided with this type of task on many occasions throughout the school year.

Every attempt should be made to determine the meaning and problem-solving strategies embedded in each student's paper. Legibility issues should be minimized.

After the students have completed this type of independent practice, it is the role of the teacher to return to the model lesson and cycle through the lesson again before the next independent practice is presented to the class.

Teachers need to look for patterns of common error in the students' work and correct the students' misconceptions during the next lesson.

Evaluation of Student Work

Teachers should not feel a need to comment on every student paper every day. The use of a rubric helps teachers provide feedback to students in a meaningful way. General feedback that does not address the student's difficulty with math (e.g., awesome, good job, needs work) does not provide adequate feedback that will help the student to improve.

Some diagnostic questions and comments that could be used include the following:

- Let's discuss different methods that were used by members of the class to solve the problem.
- Do you agree with this answer? Why or why not?
- What different methods could you use to solve the problem?
- What if you used different numbers, shapes, or figures?
- Is there more than one right answer?

Use of Rubrics to Assess Student Work

When scoring short-answer and extended-response items, rubrics should be used. There are many rubrics available for use by classroom teachers that can be applied to these problem types. Point values can vary. *Math on Target* uses a two-point rubric for short-answer items and a four-point rubric for extended-response items.

The goal of using rubrics is to assess the student's problem-solving abilities, not simply to score a student's final answer as right or wrong.

4-Point Rubric for Extended-Response Questions

<u>4 points</u>:
- Contains an effective solution using correct computation and/or problem set-up
- Shows a complete understanding of the mathematical problem
- Explains the points relevant to the solution
- Demonstrates logical reasoning and valid conclusions
- Communicates effectively using pictures, words, and/or numbers (symbols)

<u>3 points</u>:
- Contains an effective solution with minor errors in computation and/or problem set-up
- Shows an understanding of the mathematical problem
- Explains most of the points relative to the solution but neglects to address some aspects of the solution
- Demonstrates generally reasonable and valid conclusions
- Communicates adequately using pictures, words, and/or numbers (symbols)

<u>2 points</u>:
- Contains no solution or a flawed solution
- Demonstrates limited understanding of the mathematical problem
- Does not address the most relevant points of the solution
- Faulty reasoning
- Weak conclusion
- Limited or ineffective communication using pictures, words, and/or numbers (symbols)

4-Point Rubric (continued)

<u>1 point</u>:
- Contains no solution
- Demonstrates no understanding of the mathematical problem
- Does not address the solution
- Little or no reasoning
- No conclusion
- Invalid communication using pictures, words, and/or numbers (symbols)

<u>0 points</u>:
- No data
- No attempt

2-Point Rubric for Short-Answer Questions

<u>2 points</u>:
- Contains an effective solution
- Shows a complete understanding
- Shows logical reasoning
- Shows correct computation and/or problem setup

<u>1 point</u>:
- Contains a flawed solution
- Shows some understanding

<u>0 points</u>:
- Shows no understanding
- No data
- No attempt

Model Lesson: Putting It All Together, Step-by-Step

The following steps are derived from the four components of the Model Lesson. Each time you teach a lesson, you may want to refer to these steps. This model lesson guide is directly matched to the problem-solving activities in the chapters that follow.

Prior Knowledge

Step 1:

Before you begin, ask students the following questions:
- "What do you know about (insert a standard, a content topic, or a mathematical process)?"
- "Where have you seen this in your life?"

Model

Step 2:
- Explain how the thinking map should be used by completing it on an overhead or by providing each student with a personal copy.
- Introduce a model problem from *Math on Target* that illustrates the standard, content topic, or mathematical process that you want to teach.
- As a whole class activity, demonstrate the completion of the thinking map on the overhead projector using a think-aloud technique.

Model Lesson: Putting It All Together, Step-by-Step

Guided Practice

Step 3:

Introduce the problem students will be solving, as well as the thinking map.

Instruct students in completing the thinking map to solve the problem. The thinking map will help the students systematically solve the problem.

Provide ongoing support as students complete the thinking map to solve the problem or problems.

Independent Practice

Step 4:

When you determine that the students are ready for independent practice, provide them with a real-life problem from *Math on Target*. The final step is scoring the student papers. Score the papers based on the scoring rubric. Students should become familiar with the standards of the rubrics.

Preface Two

Design Your Own Higher-Level Mathematics Problem

The subject matter contained in the mathematics problem should be familiar to the students and should involve situations they understand.

The prerequisite mathematical skills needed to solve the problem should be taught prior to asking students to solve problems requiring those skills.

The mathematics problem should have meaning to the students and should incorporate something they might encounter in real life.

The directions for how you want students to show their work should be clearly described in the problem.

The evaluation criteria and scoring rubric should be reviewed with the students before they are given the mathematics problem(s). The scoring rubric should not be a mystery to the students. Students need to develop a clear understanding of the standards and scoring procedures. We have found that students become excellent scorers when trained to use the rubric.

Some questions you may want to ask yourself after you have created your problem:
- Does the problem relate to the students' lives?
- Does the problem require higher-level thinking?
- Does solving the problem require multiple steps?
- Are there various methods that could be used to arrive at a solution?
- Have the curricular skills, concepts, and processes required to solve the problem been taught?

Chapter 1

Number Sense

The purposes of this chapter include

1. Providing a definition of Number Sense and what it means for fourth-grade students.
2. Offering teaching tips on where students break down in Number Sense problems.
3. Offering test-taking tips on multiple-choice, short-answer, and extended-response questions dealing with Number Sense.
4. Providing ideas for connections with Number Sense concepts and processes with other areas of the curriculum.

The following teaching tools are provided for extending students' thinking with Number Sense:

- Model lesson for Number Sense multiple-choice problems
- Seven Number Sense multiple-choice practice problems
- Model lesson for Number Sense short-answer problems
- Two Number Sense short-answer practice problems
- Model lesson for Number Sense extended-response problems
- One Number Sense extended-response practice problem

What is Number Sense?

Number sense involves a student's ability to understand numbers and number systems and to represent numbers in a variety of ways. In grade 4, students explore equivalent forms of fractions, decimals, and whole numbers using pictures, diagrams, and real-world examples. As they develop greater fluency in computation, students discover the associative and distributive properties to solve problems. The structure of the base-ten number system is emphasized as numbers are represented in whole numbers through millions and decimals through the thousandths place. Students also represent factors and multiples of whole numbers as they determine whether numbers are prime or composite. The concepts of division with remainders is explored using money or concrete examples. Students use a variety of tools and strategies to solve multi-step problems, such as mental math, paper and pencil, and a calculator. Finally, students will analyze and interpret their answers to determine whether their results are reasonable.

© Englefield & Associates, Inc.   COPYING IS PROHIBITED

Chapter 1—Number Sense

What does Number Sense look like?

- Students identify equivalent forms of fractions, decimals, and whole numbers using numbers, words, or pictures.
- Students will understand how the base-ten system works with whole numbers through millions and decimals through thousandths.
- Students will provide the factors for numbers up to 100 and identify prime and composite numbers.
- Students will round numbers to a provided place value.
- Students will use the associative and distributive properties to solve problems.
- Students will use strategies to interpret and solve a variety of problems:
  1. Solve division problems with remainders.
  2. Count money and make change.
  3. Estimate the results of problems using whole numbers, decimals, and fractions.
  4. Perform mental computations.
  5. Solve multi-step problems.
  6. Use a variety of tools—mental math, paper and pencil, and a calculator.
  7. Explain and interpret their results for accuracy, logic, and reasonableness.

Vocabulary related to Number Sense

Associative properties: In addition and multiplication, the answer will be the same no matter how the numbers are grouped. For example, 5 + 2 + 3 = 10, (5 + 2) + 3 = 10, 5 + (2 + 3) = 10.

Common denominator: A common multiple, or the number into which the denominators of fractions will divide exactly.

Commutative property: A property that applies to adding and multiplying numbers. Numbers can be added together or multiplied together in any order. For example, 7 + 3 = 10 is the same as 3 + 7 = 10; 6 x 2 = 12 is the same as 2 x 6 = 12.

Decimals: Numbers that are expressed as tenths, hundredths, and thousandths (e.g., 3 and 8 tenths as 3.8).

Distributive property: The product of a number and the sum (or difference) of two numbers is equal to the sum (or difference) of the two products. For example, 5 x 9 is the same as 5 x (4 + 5) and (5 x 4) + (5 x 5) = 20 + 25 = 45.

Estimation: An approximate calculation; Sally has $42.36, or approximately $40.00, in her bank.

Estimation strategies: Methods used to make approximate calculations, often based on rounding off. Sam's candy cost $4.08, or approximately $4.00. Twelve boxes cost $4.08 x 12, or approximately $48.00.

Equivalent: Having the same value, such as equivalent fractions (2/4 and 1/2 are equivalent fractions).

Equivalence of fractions: Having the same value or amount. Two halves are equivalent to a whole.

Fractions: Numbers that represent parts in sets (1/2, 1/3, or 1/6).

Numbers less than zero: Negative numbers; for example, -20.

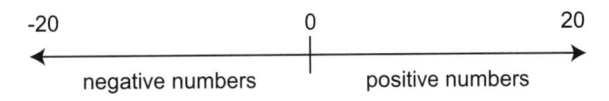

```
      -20              0              20
       <---------------|--------------->
       negative numbers   positive numbers
```

Number line: A line marked with numbers.

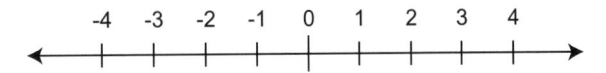

```
   -4  -3  -2  -1  0  1  2  3  4
   <---|---|---|---|---|---|---|---|--->
```

Percent: Number out of 100; uses % as a symbol. For example, 75/100 = 75 out of 100 = 0.75 = 75%.

Place value: The value of a digit that is determined by its place in a number. For example, the 1 is in the ones place in the number 991; the 1 is in the tens place in the number 919; and the 1 is in the hundreds place in the number 199.

Teaching Tips:

- Provide students with physical, verbal, and graphic representations of fractions, decimals, and whole numbers. For example:

    2/6 = 1/3 = two-sixths = 0.33 = ▨□□ = ▨□□□□□

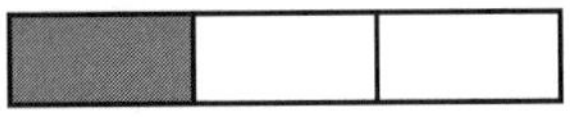

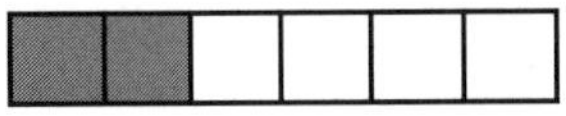

- Provide students with problems that show the meaning of remainders in division problems, such as the following:

    There are 43 students and 45 party favor bags. If each child receives a party favor bag, how many bags will be left? (2)

- Set up a class store and have students make change using coins or paper bills.

- Area models can provide students with a visual representation of how fractions are related to the whole or are part of a whole.

- Provide a variety of activities for fractions and decimals that focus on equivalence so that students can easily move among the three mathematical representations. (For example, 3/4 is equivalent to 6/8, and its decimal equivalent is 0.75.)

- Provide practice with common percents such as 10%, 33 1/3%, or 50%, using problems involving the purchase of sale items. This is a perfect activity to use to practice solving mental mathematics problems.

- Use temperature to introduce the concept of negative numbers.

Chapter 1—Number Sense

Model Problem 1: Number Sense Multiple-Choice

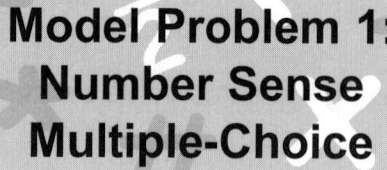

1. Prior Knowledge

Determine the student's prior knowledge about the strand, content of the problem, and/or mathematical process of the problem.

These questions can be used to help determine students' knowledge:

- What do you know about changing fractions to decimals?
- Can you think of a time when you have used decimals instead of fractions?

Additional prompts could be provided by using the following topics:

- food
- money
- cooking
- shopping

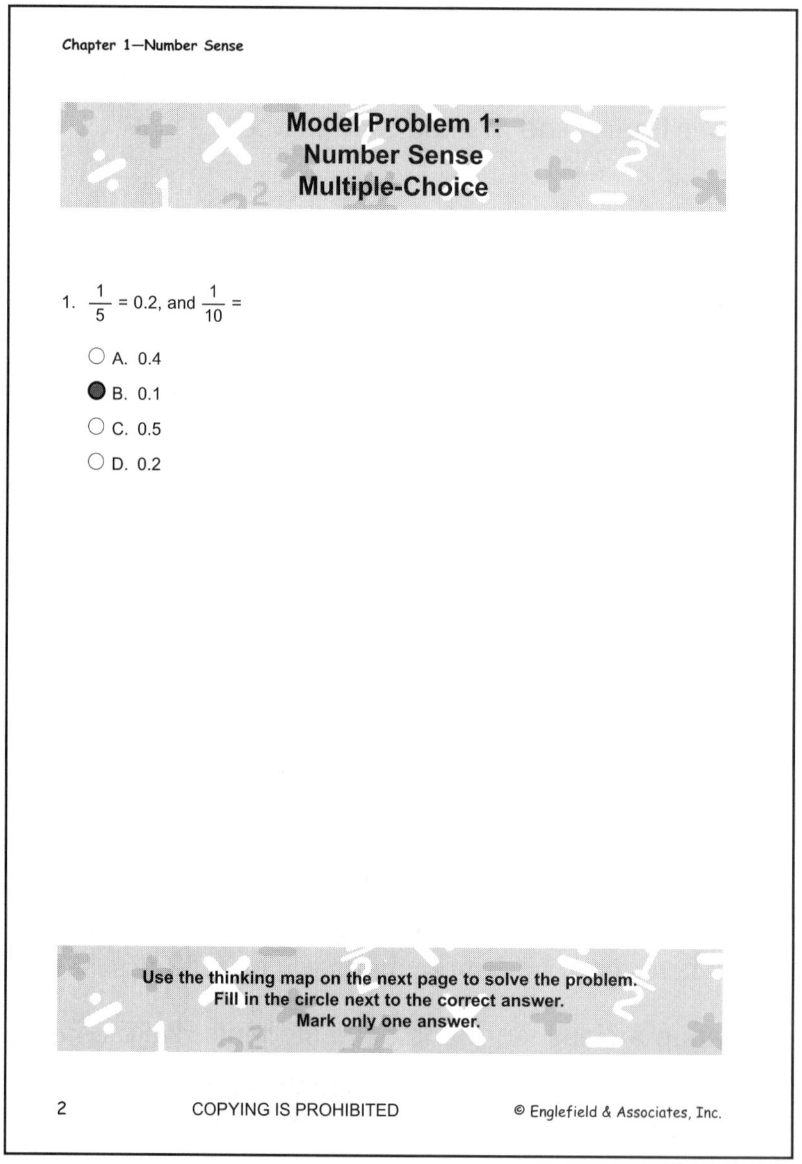

Chapter 1—Number Sense

Model Problem 1: Number Sense Multiple-Choice

1. $\frac{1}{5} = 0.2$, and $\frac{1}{10} =$
   - ○ A. 0.4
   - ● B. 0.1
   - ○ C. 0.5
   - ○ D. 0.2

Use the thinking map on the next page to solve the problem.
Fill in the circle next to the correct answer.
Mark only one answer.

Chapter 1—Number Sense

2. Model

Model the use of the mathematical thinking map through class discussion with the thinking process modeled aloud as the thinking map is completed.

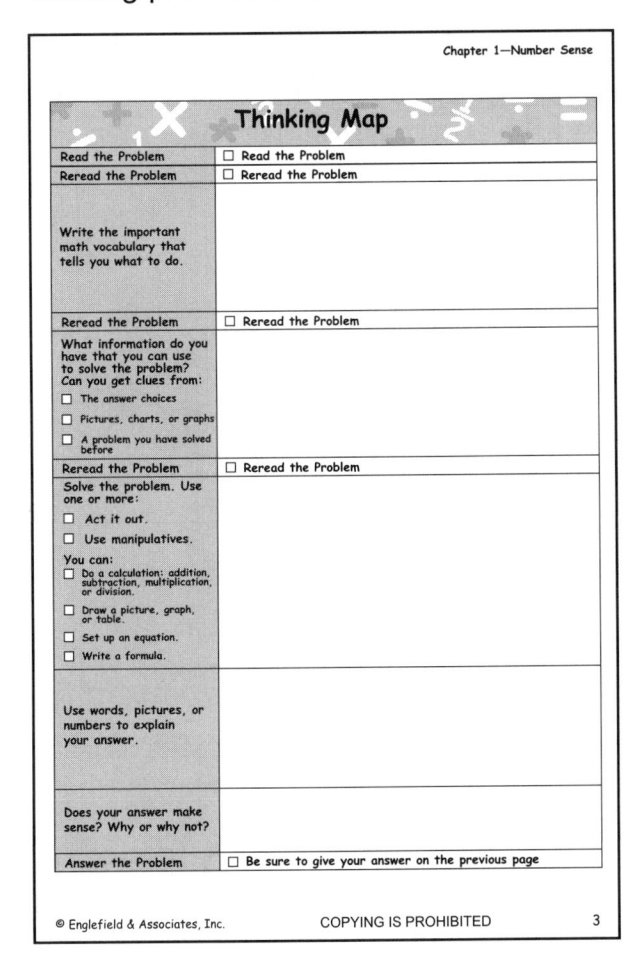

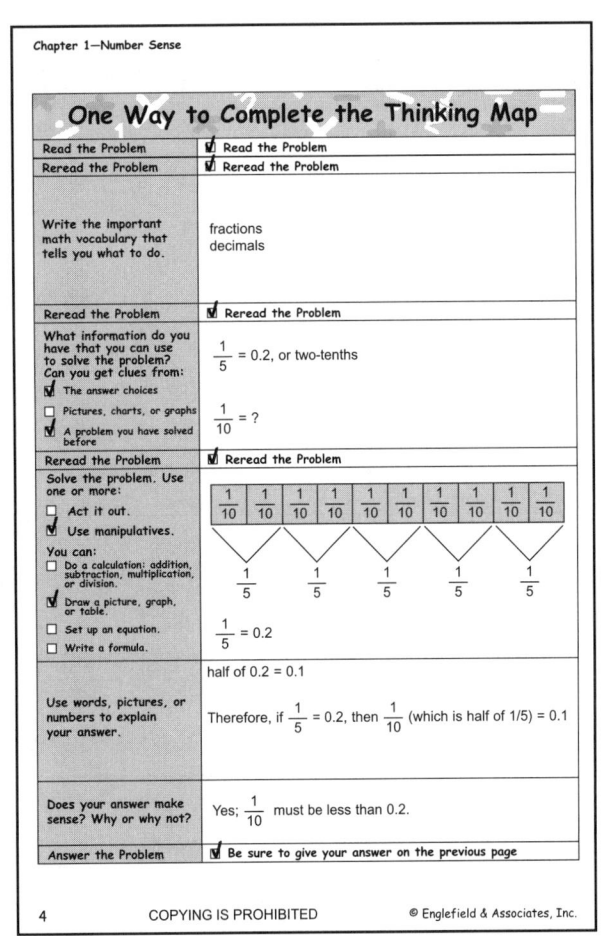

Thinking Map	
Read the Problem	☐ Read the Problem
Reread the Problem	☐ Reread the Problem
Write the important math vocabulary that tells you what to do.	
Reread the Problem	☐ Reread the Problem
What information do you have that you can use to solve the problem? Can you get clues from: ☐ The answer choices ☐ Pictures, charts, or graphs ☐ A problem you have solved before	
Reread the Problem	☐ Reread the Problem
Solve the problem. Use one or more: ☐ Act it out. ☐ Use manipulatives. You can: ☐ Do a calculation: addition, subtraction, multiplication, or division. ☐ Draw a picture, graph, or table. ☐ Set up an equation. ☐ Write a formula.	
Use words, pictures, or numbers to explain your answer.	
Does your answer make sense? Why or why not?	
Answer the Problem	☐ Be sure to give your answer on the previous page

One Way to Complete the Thinking Map	
Read the Problem	☑ Read the Problem
Reread the Problem	☑ Reread the Problem
Write the important math vocabulary that tells you what to do.	fractions decimals
Reread the Problem	☑ Reread the Problem
What information do you have that you can use to solve the problem? Can you get clues from: ☐ The answer choices ☐ Pictures, charts, or graphs ☑ A problem you have solved before	$\frac{1}{5}$ = 0.2, or two-tenths $\frac{1}{10}$ = ?
Reread the Problem	☑ Reread the Problem
Solve the problem. Use one or more: ☐ Act it out. ☑ Use manipulatives. You can: ☐ Do a calculation: addition, subtraction, multiplication, or division. ☑ Draw a picture, graph, or table. ☐ Set up an equation. ☐ Write a formula.	[diagram showing ten $\frac{1}{10}$ segments grouped in pairs equaling $\frac{1}{5}$] $\frac{1}{5}$ = 0.2
Use words, pictures, or numbers to explain your answer.	half of 0.2 = 0.1 Therefore, if $\frac{1}{5}$ = 0.2, then $\frac{1}{10}$ (which is half of 1/5) = 0.1
Does your answer make sense? Why or why not?	Yes; $\frac{1}{10}$ must be less than 0.2.
Answer the Problem	☑ Be sure to give your answer on the previous page

3. Guided Practice

Provide guided practice by presenting a problem for students to complete that matches the strand, content, or process you want to teach.

4. Independent Practice

Provide independent practice so students are provided with multiple-choice, short-answer, or extended-response problems.

Chapter 1—Number Sense

Thinking Map

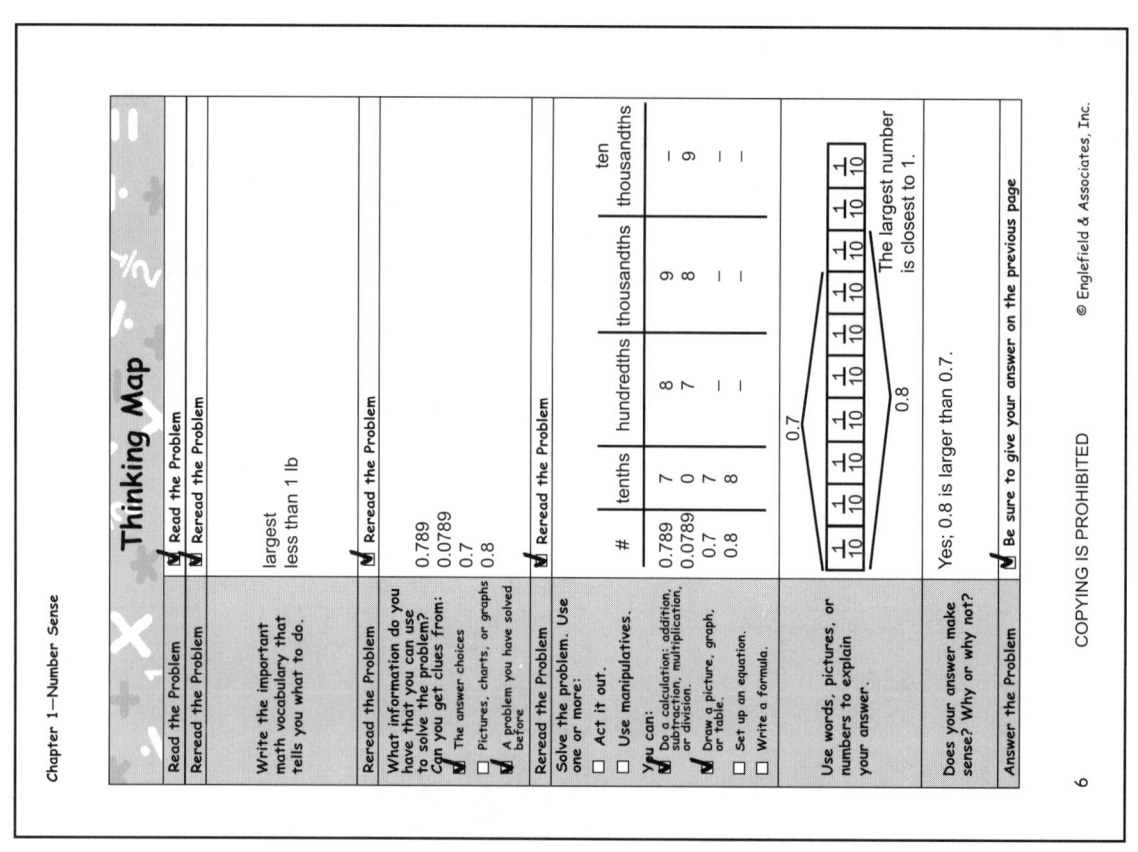

Read the Problem	☑ Read the Problem
Reread the Problem	☑ Reread the Problem
Write the important math vocabulary that tells you what to do.	largest less than 1 lb
Reread the Problem	☑ Reread the Problem
What information do you have that you can use to solve the problem? Can you get clues from: ☑ The answer choices ☐ Pictures, charts, or graphs ☐ A problem you have solved before	0.789 0.0789 0.7 0.8
Reread the Problem	☑ Reread the Problem
Solve the problem. Use one or more: ☐ Act it out. ☐ Use manipulatives. You can: ☑ Do a calculation: addition, subtraction, multiplication, or division. ☑ Draw a picture, graph, or table. ☐ Set up an equation. ☐ Write a formula.	# | tenths | hundredths | thousandths | ten thousandths 0.789 | 7 | 8 | 9 | — 0.0789 | 0 | 7 | 8 | 9 0.7 | 7 | — | — | — 0.8 | 8 | — | — | — $\frac{1}{10}\frac{1}{10}\frac{1}{10}\frac{1}{10}\frac{1}{10}\frac{1}{10}\frac{1}{10}\frac{1}{10}\frac{1}{10}\frac{1}{10}$ 0.7 0.8 The largest number is closest to 1.
Use words, pictures, or numbers to explain your answer.	
Does your answer make sense? Why or why not?	Yes; 0.8 is larger than 0.7.
Answer the Problem	☑ Be sure to give your answer on the previous page

In the Student Workbook, the Thinking Map is blank.
In the Parent/Teacher Edition, suggested ways for completing the Thinking Maps are presented.

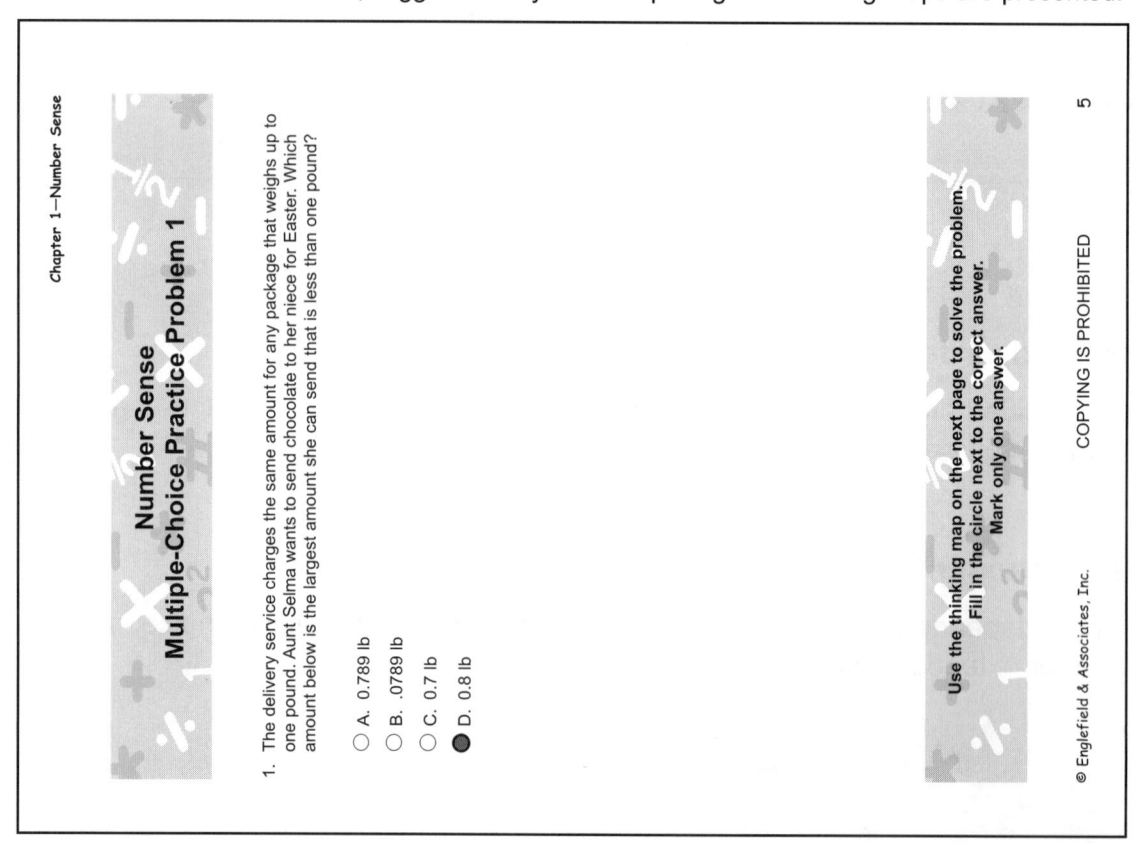

Number Sense
Multiple-Choice Practice Problem 1

1. The delivery service charges the same amount for any package that weighs up to one pound. Aunt Selma wants to send chocolate to her niece for Easter. Which amount below is the largest amount she can send that is less than one pound?

   ○ A. 0.789 lb
   ○ B. .0789 lb
   ○ C. 0.7 lb
   ● D. 0.8 lb

Use the thinking map on the next page to solve the problem.
Fill in the circle next to the correct answer.
Mark only one answer.

Thinking Map

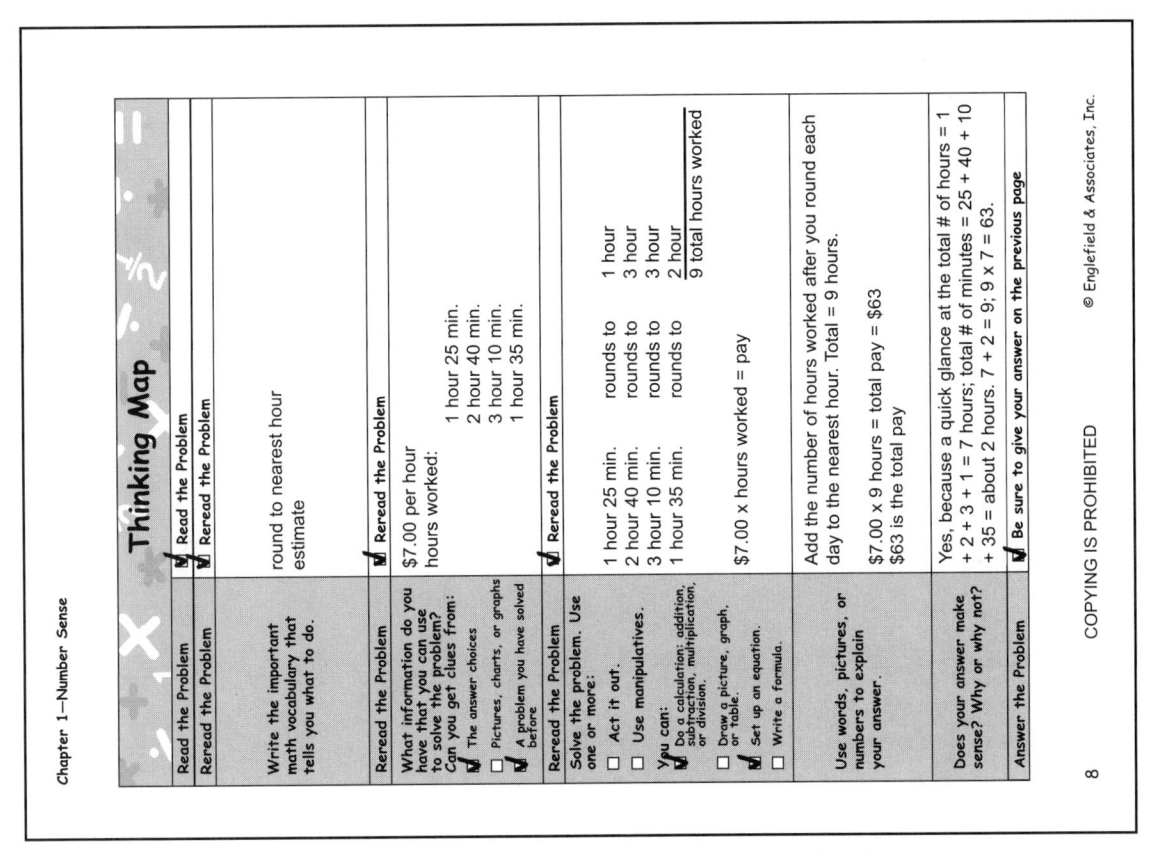

Read the Problem	
☑ Read the Problem	
☑ Reread the Problem	
Write the important math vocabulary that tells you what to do.	round to nearest hour estimate
Reread the Problem	
What information do you have that you can use to solve the problem? Can you get clues from: ☑ The answer choices ☐ Pictures, charts, or graphs ☑ A problem you have solved before	$7.00 per hour hours worked: 1 hour 25 min. 2 hour 40 min. 3 hour 10 min. 1 hour 35 min.
Reread the Problem	
Solve the problem. Use one or more: ☐ Act it out. ☐ Use manipulatives. You can: ☑ Do a calculation: addition, subtraction, multiplication, or division. ☐ Draw a picture, graph, or table. ☑ Set up an equation. ☐ Write a formula.	1 hour 25 min. rounds to 1 hour 2 hour 40 min. rounds to 3 hour 3 hour 10 min. rounds to 3 hour 1 hour 35 min. rounds to 2 hour 9 total hours worked $7.00 x hours worked = pay
Use words, pictures, or numbers to explain your answer.	Add the number of hours worked after you round each day to the nearest hour. Total = 9 hours. $7.00 x 9 hours = total pay = $63 $63 is the total pay
Does your answer make sense? Why or why not?	Yes, because a quick glance at the total # of hours = 1 + 2 + 3 + 1 = 7 hours; total # of minutes = 25 + 40 + 10 + 35 = about 2 hours. 7 + 2 = 9; 9 x 7 = 63.
Answer the Problem	
☑ Be sure to give your answer on the previous page	

In the Student Workbook, the Thinking Map is blank.
In the Parent/Teacher Edition, suggested ways for completing the Thinking Maps are presented.

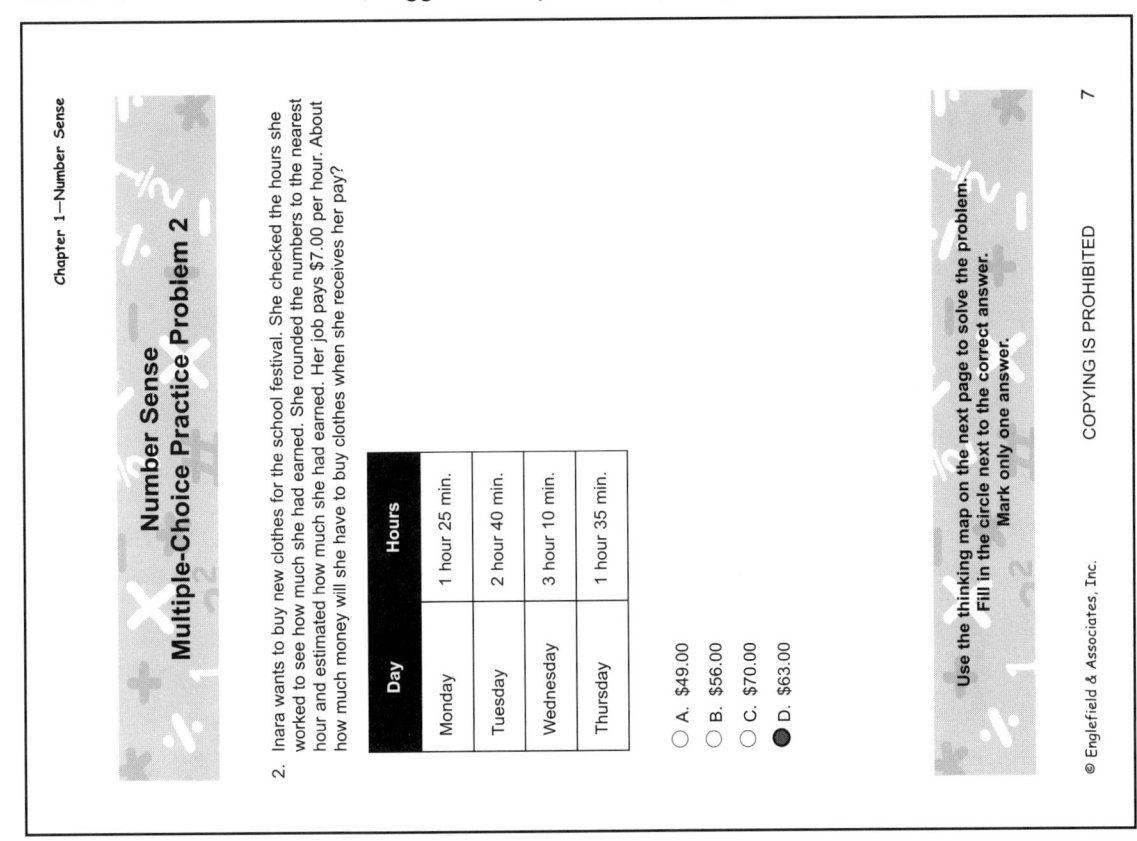

Number Sense
Multiple-Choice Practice Problem 2

2. Inara wants to buy new clothes for the school festival. She checked the hours she worked to see how much she had earned. She rounded the numbers to the nearest hour and estimated how much she had earned. Her job pays $7.00 per hour. About how much money will she have to buy clothes when she receives her pay?

Day	Hours
Monday	1 hour 25 min.
Tuesday	2 hour 40 min.
Wednesday	3 hour 10 min.
Thursday	1 hour 35 min.

○ A. $49.00
○ B. $56.00
○ C. $70.00
● D. $63.00

Use the thinking map on the next page to solve the problem.
Fill in the circle next to the correct answer.
Mark only one answer.

Thinking Map

Read the Problem ☑	
Reread the Problem ☑	multiplication property
Reread the Problem ☑	multiply by 0 property commutative property identity property distributive property
Reread the Problem ☑	By process of elimination: multiply by 0 property = a x 0 = 0 commutative property = a x b = b x a identity property = a x 1 = a None of these properties apply. The distributive property allows you to regroup an operation.
Write the important math vocabulary that tells you what to do.	
What information do you have that you can use to solve the problem? Can you get clues from: ☑ The answer choices ☐ Pictures, charts, or graphs ☑ A problem you have solved before	
Solve the problem. Use one or more: ☐ Act it out. ☐ Use manipulatives. You can: ☑ Do a calculation: addition, subtraction, multiplication, or division. ☐ Draw a picture, graph, or table. ☐ Set up an equation. ☐ Write a formula.	How many seconds in 120 minutes = multiply 120 x 60 Distributive property: 120 x (6 x 10) = 120 x 60 = 720 x 10 = 7200
Use words, pictures, or numbers to explain your answer.	
Does your answer make sense? Why or why not?	Yes, by process of elimination, it's the only one that works.
Answer the Problem ☑	Be sure to give your answer on the previous page

10   COPYING IS PROHIBITED   © Englefield & Associates, Inc.

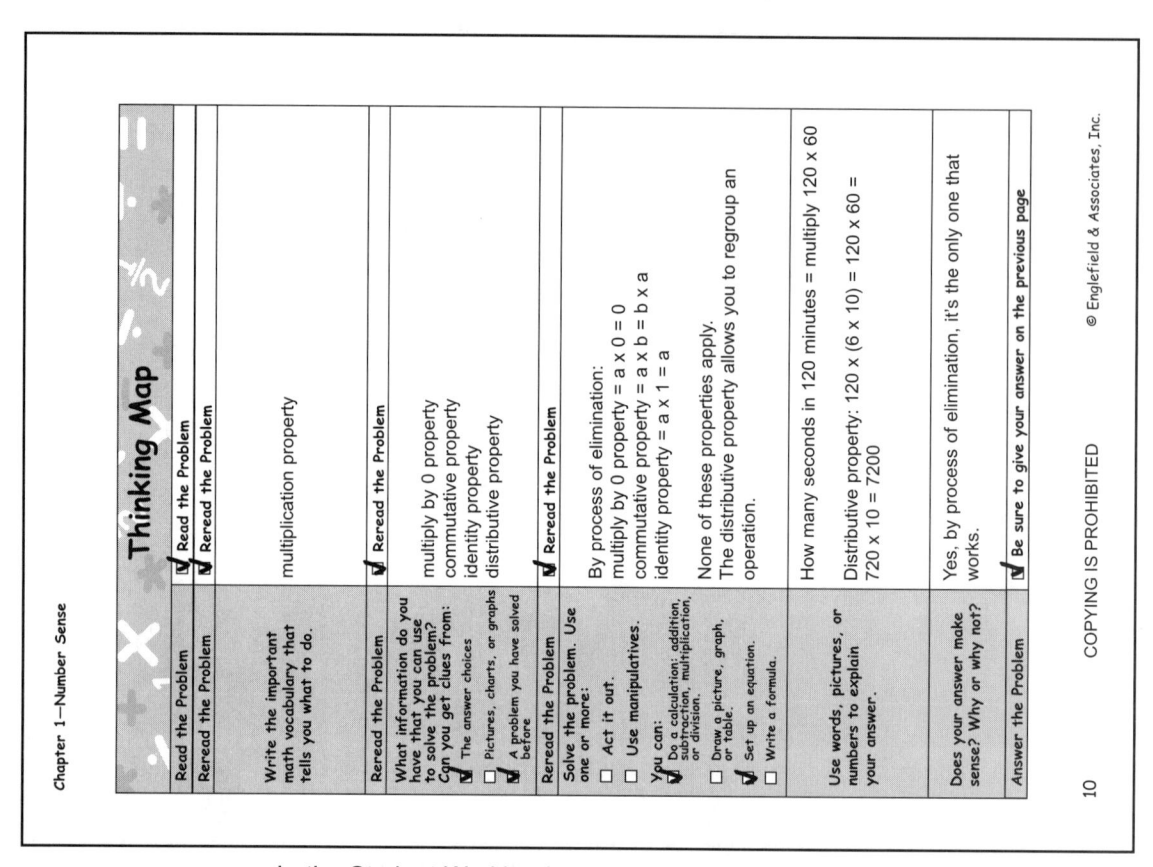

In the Student Workbook, the Thinking Map is blank.
In the Parent/Teacher Edition, suggested ways for completing the Thinking Maps are presented.

Chapter 1—Number Sense

Number Sense
Multiple-Choice Practice Problem 3

3. Which property of multiplication would you use to solve the following problem in your head?

   How many seconds are in 120 minutes?

   ○ A. Multiply by 0 property
   ○ B. Commutative property
   ● C. Distributive property
   ○ D. Identity property

Use the thinking map on the next page to solve the problem.
Fill in the circle next to the correct answer.
Mark only one answer.

© Englefield & Associates, Inc.   COPYING IS PROHIBITED   9

Chapter 1—Number Sense

Thinking Map

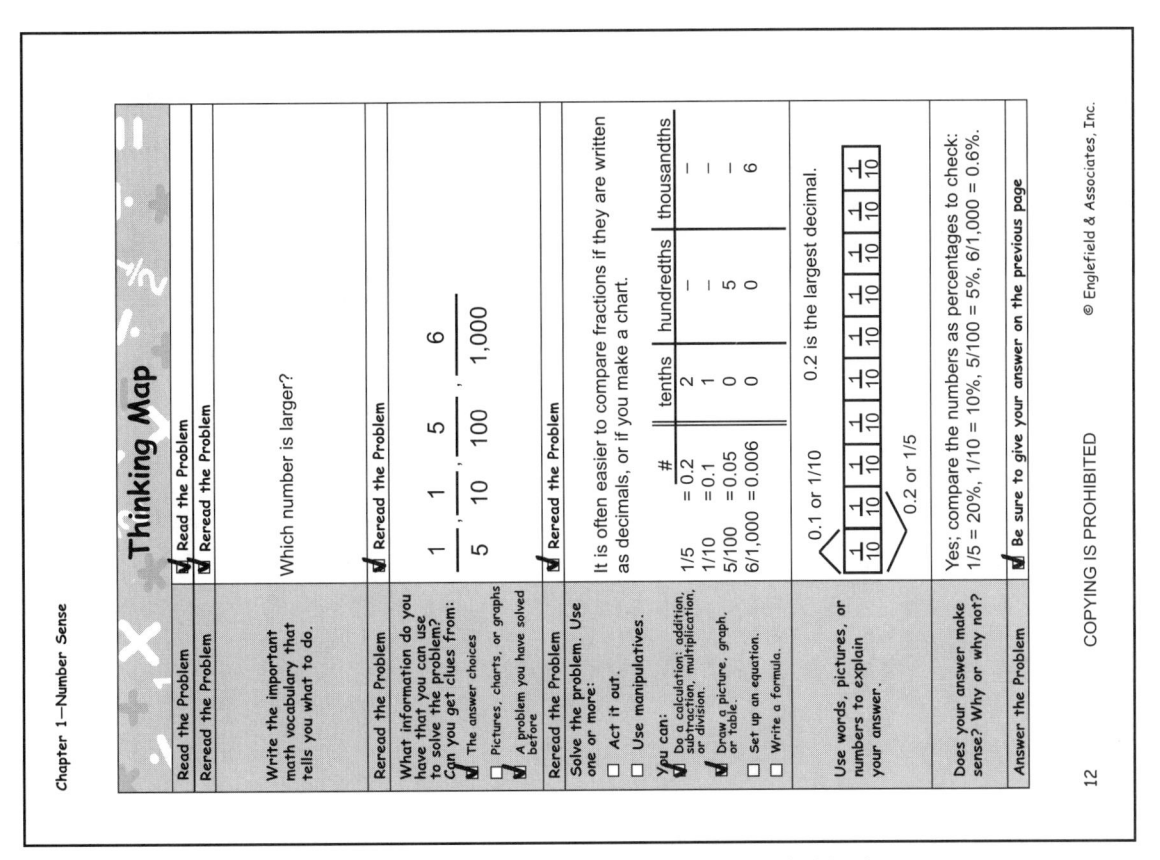

Read the Problem	☑ Read the Problem ☑ Reread the Problem
Write the important math vocabulary that tells you what to do.	Which number is larger?
Reread the Problem	☑ Reread the Problem
What information do you have that you can use to solve the problem? Can you get clues from: ☐ The answer choices ☑ Pictures, charts, or graphs ☐ A problem you have solved before	$\frac{1}{5}, \frac{1}{10}, \frac{5}{100}, \frac{6}{1,000}$
Reread the Problem	☑ Reread the Problem
Solve the problem. Use one or more: ☐ Act it out. ☐ Use manipulatives. You can: ☐ Do a calculation: addition, subtraction, multiplication, or division. ☑ Draw a picture, graph, or table. ☐ Set up an equation. ☐ Write a formula.	It is often easier to compare fractions if they are written as decimals, or if you make a chart. | # | tenths | hundredths | thousandths | |---|---|---|---| | 1/5 = 0.2 | 2 | — | — | | 1/10 = 0.1 | 1 | — | — | | 5/100 = 0.05 | 0 | 5 | — | | 6/1,000 = 0.006 | 0 | 0 | 6 | 0.2 is the largest decimal. $\frac{1}{10} \frac{1}{10} \frac{1}{10} \frac{1}{10} \frac{1}{10} \frac{1}{10} \frac{1}{10} \frac{1}{10} \frac{1}{10} \frac{1}{10}$ 0.1 or 1/10 > 0.2 or 1/5
Use words, pictures, or numbers to explain your answer.	
Does your answer make sense? Why or why not?	Yes; compare the numbers as percentages to check: 1/5 = 20%, 1/10 = 10%, 5/100 = 5%, 6/1,000 = 0.6%.
Answer the Problem	☑ Be sure to give your answer on the previous page

In the Student Workbook, the Thinking Map is blank.
In the Parent/Teacher Edition, suggested ways for completing the Thinking Maps are presented.

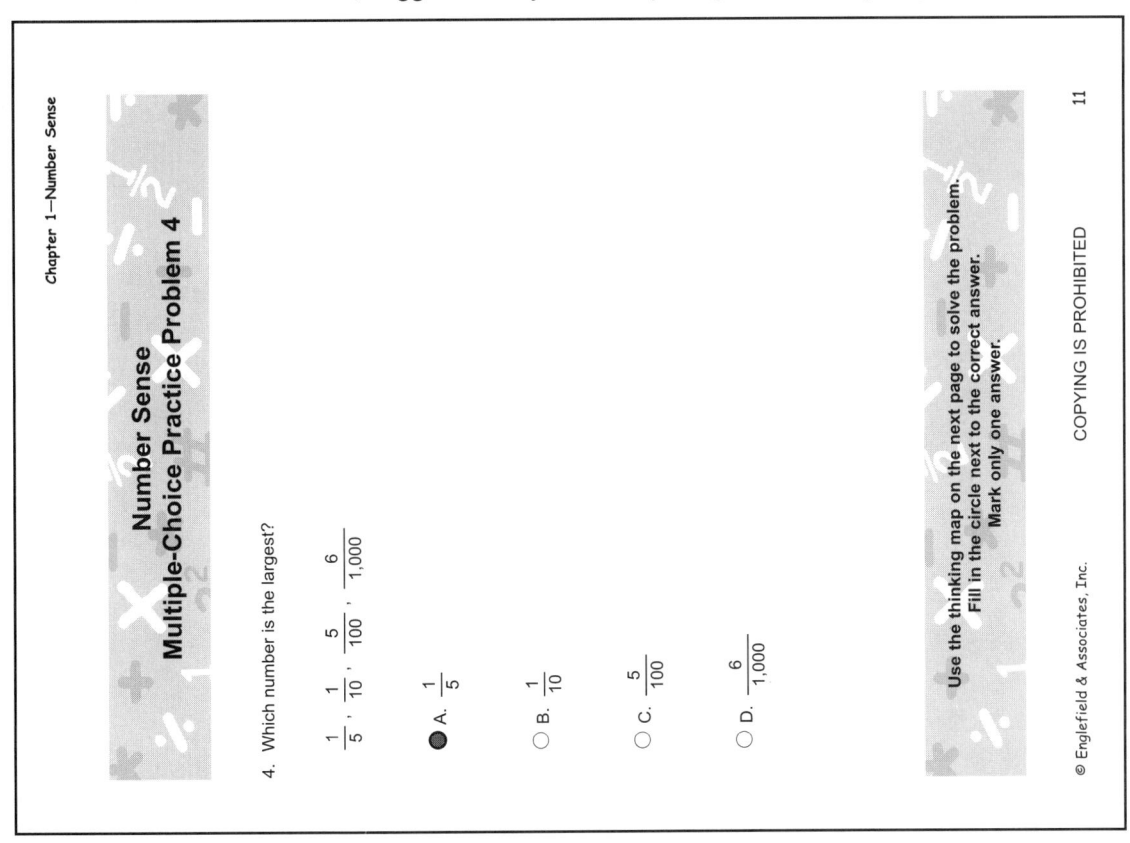

Number Sense
Multiple-Choice Practice Problem 4

4. Which number is the largest?

$$\frac{1}{5}, \frac{1}{10}, \frac{5}{100}, \frac{6}{1,000}$$

● A. $\frac{1}{5}$

○ B. $\frac{1}{10}$

○ C. $\frac{5}{100}$

○ D. $\frac{6}{1,000}$

Use the thinking map on the next page to solve the problem.
Fill in the circle next to the correct answer.
Mark only one answer.

Chapter 1—Number Sense

Thinking Map

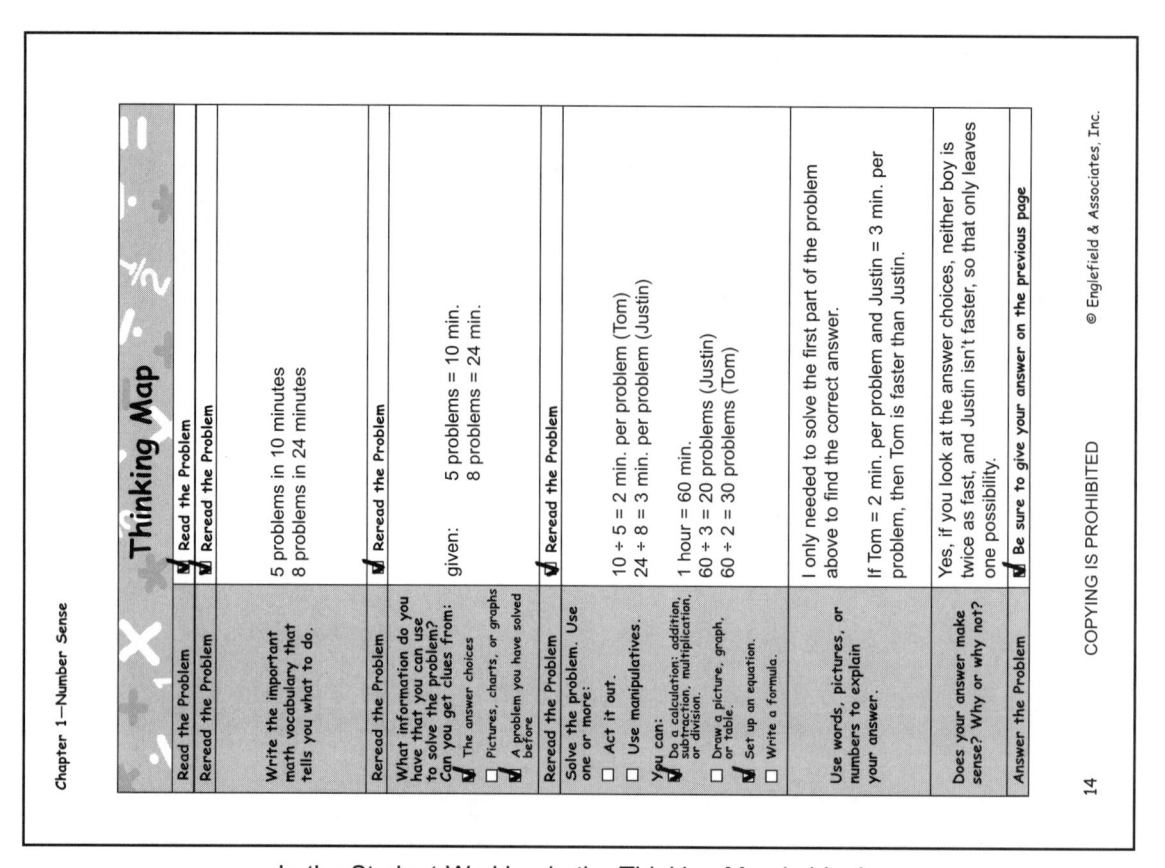

Read the Problem	☑ Read the Problem
Reread the Problem	☑ Reread the Problem
Write the important math vocabulary that tells you what to do.	5 problems in 10 minutes 8 problems in 24 minutes
Reread the Problem	☑ Reread the Problem
What information do you have that you can use to solve the problem? Can you get clues from: ☑ The answer choices ☐ Pictures, charts, or graphs ☑ A problem you have solved before	given: 5 problems = 10 min. 8 problems = 24 min.
Reread the Problem	☑ Reread the Problem
Solve the problem. Use one or more: ☐ Act it out. ☐ Use manipulatives. You can: ☑ Do a calculation: addition, subtraction, multiplication, or division. ☐ Draw a picture, graph, or table. ☑ Set up an equation. ☐ Write a formula.	$10 \div 5$ = 2 min. per problem (Tom) $24 \div 8$ = 3 min. per problem (Justin) 1 hour = 60 min. $60 \div 3$ = 20 problems (Justin) $60 \div 2$ = 30 problems (Tom)
Use words, pictures, or numbers to explain your answer.	I only needed to solve the first part of the problem above to find the correct answer. If Tom = 2 min. per problem and Justin = 3 min. per problem, then Tom is faster than Justin.
Does your answer make sense? Why or why not?	Yes, if you look at the answer choices, neither boy is twice as fast, and Justin isn't faster, so that only leaves one possibility.
Answer the Problem	☑ Be sure to give your answer on the previous page

14

In the Student Workbook, the Thinking Map is blank.
In the Parent/Teacher Edition, suggested ways for completing the Thinking Maps are presented.

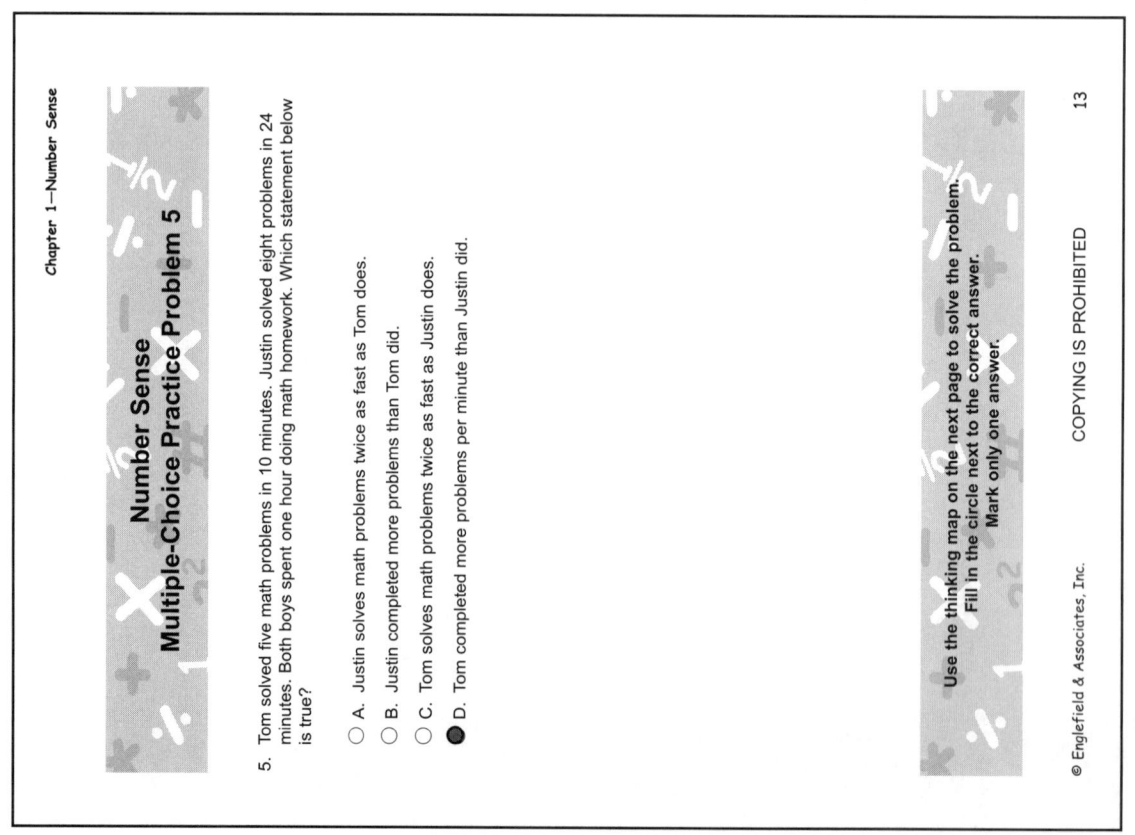

Number Sense
Multiple-Choice Practice Problem 5

5. Tom solved five math problems in 10 minutes. Justin solved eight problems in 24 minutes. Both boys spent one hour doing math homework. Which statement below is true?

   ○ A. Justin solves math problems twice as fast as Tom does.
   ○ B. Justin completed more problems than Tom did.
   ○ C. Tom solves math problems twice as fast as Justin does.
   ● D. Tom completed more problems per minute than Justin did.

Use the thinking map on the next page to solve the problem.
Fill in the circle next to the correct answer.
Mark only one answer.

13

26   COPYING IS PROHIBITED   © Englefield & Associates, Inc.

Thinking Map

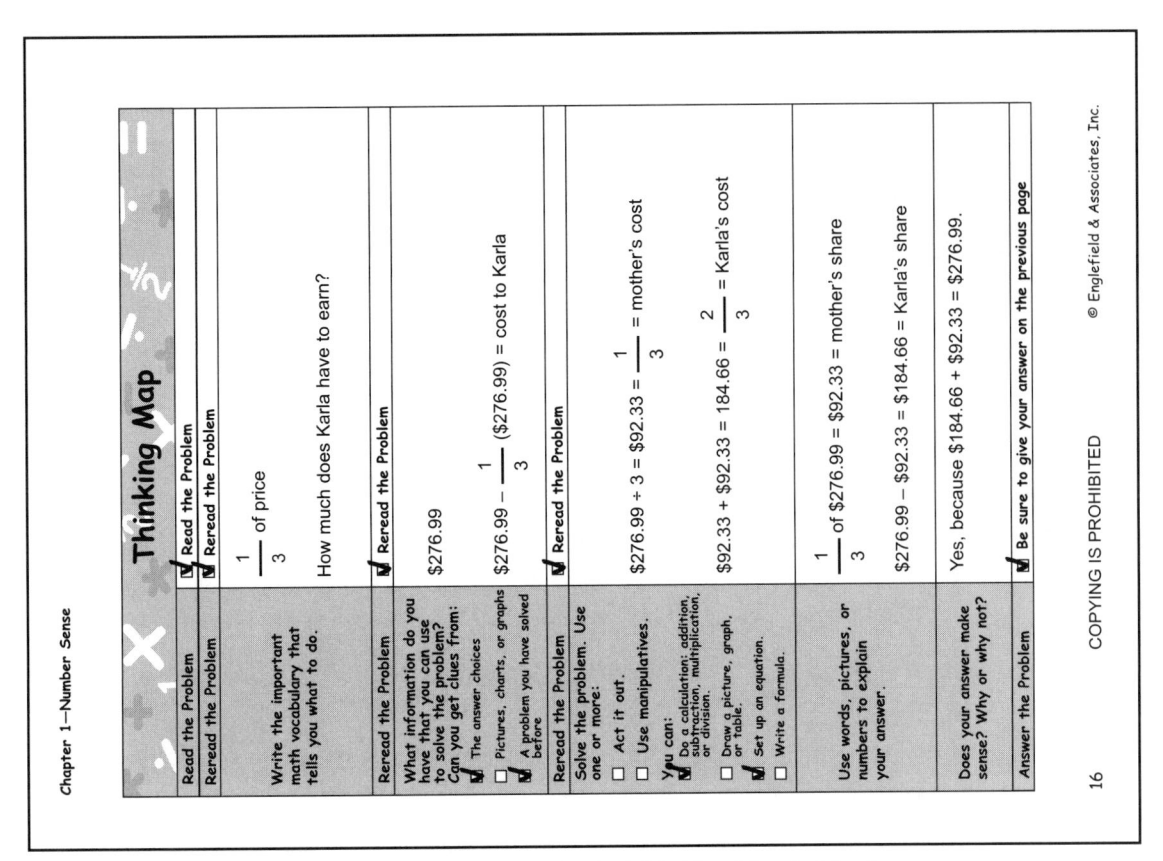

Read the Problem	☑ Read the Problem
Reread the Problem	☑ Reread the Problem
Write the important math vocabulary that tells you what to do.	$\frac{1}{3}$ of price How much does Karla have to earn?
Reread the Problem	☑ Reread the Problem
What information do you have that you can use to solve the problem? Can you get clues from: ☐ The answer choices ☑ Pictures, charts, or graphs ☐ A problem you have solved before	$276.99
Reread the Problem	☑ Reread the Problem
Solve the problem. Use one or more: ☐ Act it out. ☐ Use manipulatives. You can: ☐ Do a calculation: addition, subtraction, multiplication, or division. ☐ Draw a picture, graph, or table. ☑ Set up an equation. ☐ Write a formula.	$276.99 − $\frac{1}{3}$ ($276.99) = cost to Karla $276.99 ÷ 3 = $92.33 = $\frac{1}{3}$ = mother's cost $92.33 + $92.33 = 184.66 = $\frac{2}{3}$ = Karla's cost
Use words, pictures, or numbers to explain your answer.	$\frac{1}{3}$ of $276.99 = $92.33 = mother's share $276.99 − $92.33 = $184.66 = Karla's share
Does your answer make sense? Why or why not?	Yes, because $184.66 + $92.33 = $276.99.
Answer the Problem	☑ Be sure to give your answer on the previous page

In the Student Workbook, the Thinking Map is blank.
In the Parent/Teacher Edition, suggested ways for completing the Thinking Maps are presented.

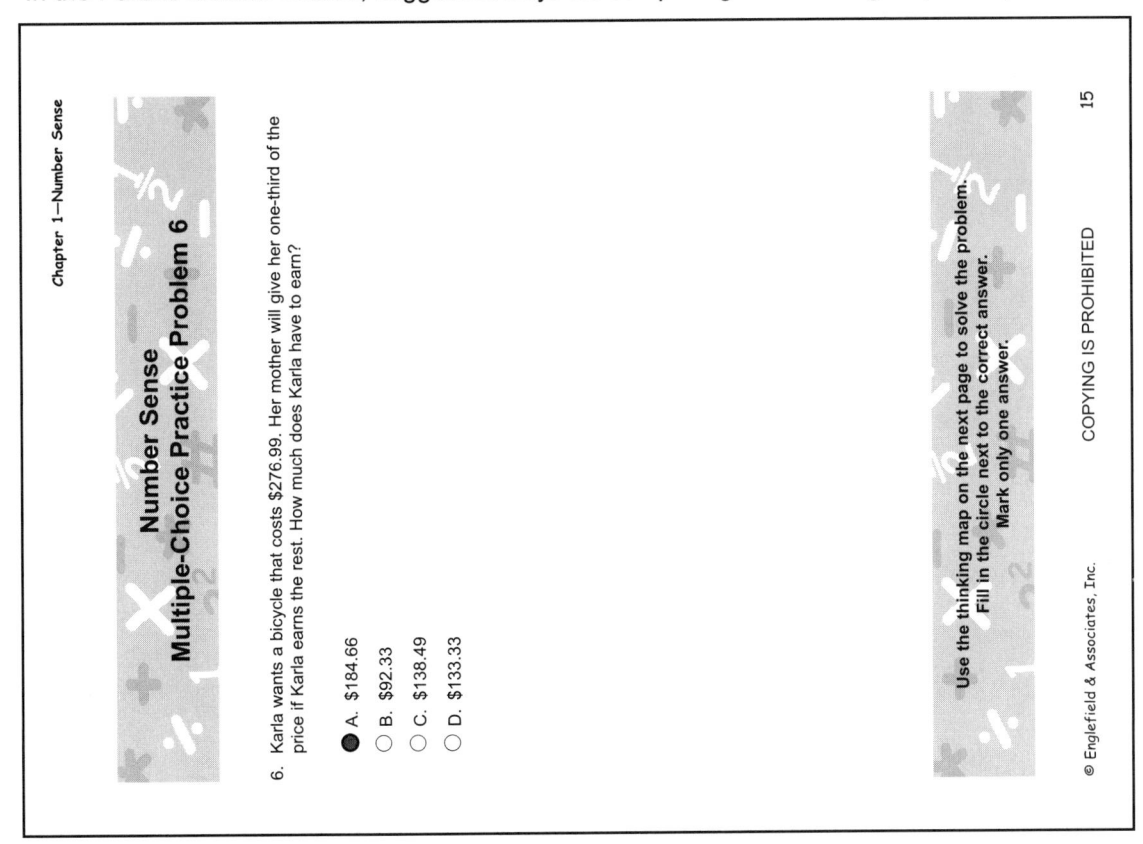

Number Sense
Multiple-Choice Practice Problem 6

6. Karla wants a bicycle that costs $276.99. Her mother will give her one-third of the price if Karla earns the rest. How much does Karla have to earn?

   ● A. $184.66
   ○ B. $92.33
   ○ C. $138.49
   ○ D. $133.33

Use the thinking map on the next page to solve the problem.
Fill in the circle next to the correct answer.
Mark only one answer.

Chapter 1—Number Sense

Thinking Map

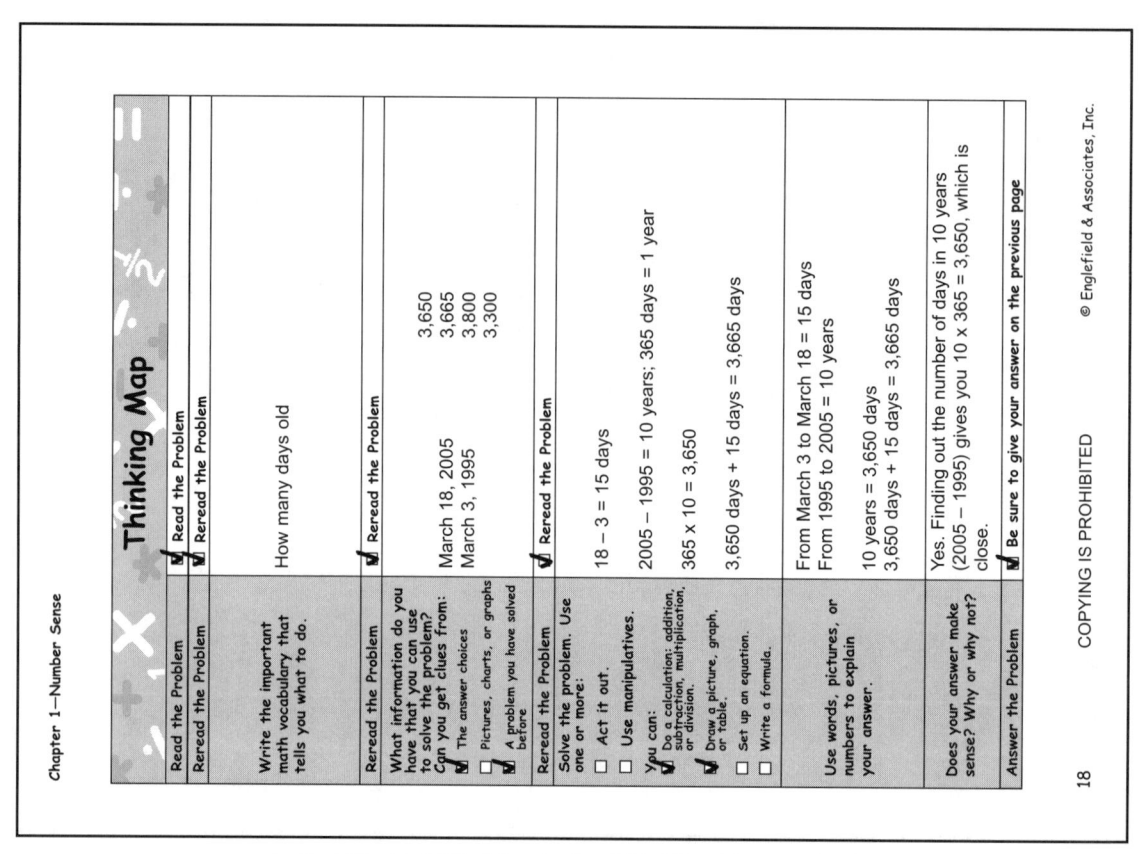

Read the Problem	☑ Read the Problem
	☑ Reread the Problem
Write the important math vocabulary that tells you what to do.	How many days old
Reread the Problem	☑ Reread the Problem

What information do you have that you can use to solve the problem? Can you get clues from:
☑ The answer choices
☐ Pictures, charts, or graphs
☑ A problem you have solved before
	March 18, 2005 3,650
	March 3, 1995 3,665
	                      3,800
	                      3,300

Reread the Problem Solve the problem. Use one or more: ☐ Act it out. ☐ Use manipulatives. You can: ☑ Do a calculation: addition, subtraction, multiplication, or division. ☑ Draw a picture, graph, or table. ☐ Set up an equation. ☐ Write a formula.	☑ Reread the Problem 18 − 3 = 15 days 2005 − 1995 = 10 years; 365 days = 1 year 365 × 10 = 3,650 3,650 days + 15 days = 3,665 days
Use words, pictures, or numbers to explain your answer.	From March 3 to March 18 = 15 days From 1995 to 2005 = 10 years 10 years = 3,650 days 3,650 days + 15 days = 3,665 days
Does your answer make sense? Why or why not?	Yes. Finding out the number of days in 10 years (2005 − 1995) gives you 10 × 365 = 3,650, which is close.
Answer the Problem	☑ Be sure to give your answer on the previous page

18

In the Student Workbook, the Thinking Map is blank.
In the Parent/Teacher Edition, suggested ways for completing the Thinking Maps are presented.

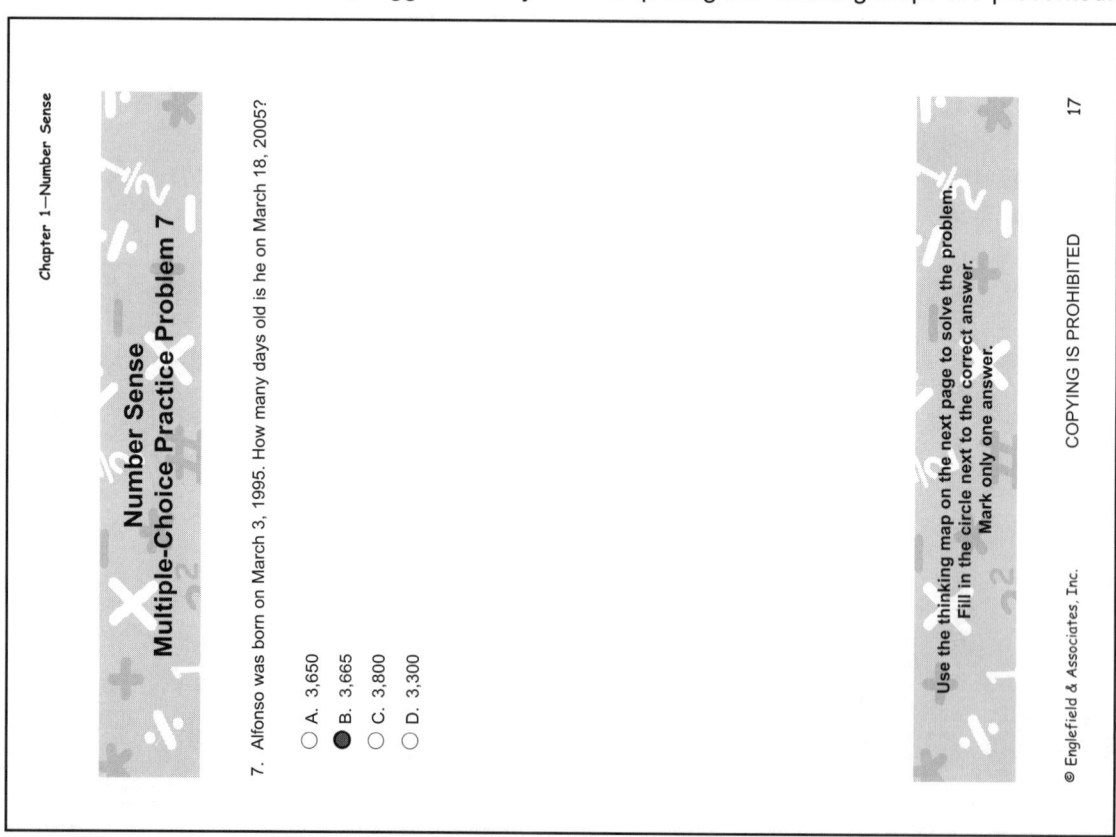

Number Sense
Multiple-Choice Practice Problem 7

7. Alfonso was born on March 3, 1995. How many days old is he on March 18, 2005?

○ A. 3,650
● B. 3,665
○ C. 3,800
○ D. 3,300

Use the thinking map on the next page to solve the problem.
Fill in the circle next to the correct answer.
Mark only one answer.

17

28   COPYING IS PROHIBITED   © Englefield & Associates, Inc.

Model Problem 2: Number Sense Short-Answer

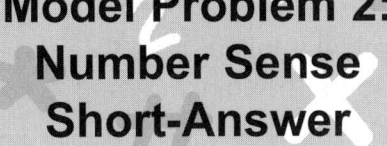

1. Prior Knowledge

Determine the student's prior knowledge about the strand, content of the problem, and/or mathematical process of the problem.

These questions can be used to help determine students' knowledge:

- What is the "order of operations"?
- When have you known the answer to a problem and worked backward to solve that problem?

Additional prompts could be provided by using the following topics:

- placing students in a school bus
- making party bags for a group event
- sharing with friends
- dividing students into teams

Model Problem 2: Number Sense Short-Answer

2. Fill in the correct operations. Explain your work.

(4 ☐ 8) ☐ 6 = 72

> 1 point: The student demonstrates understanding of the order of operations in solving the problem.
>
> 1 point: The student finds the correct answer: (4 + 8) × 6 = 72.

Use the thinking map on the next page to solve the problem. Write your answer in the box.

Chapter 1—Number Sense

2. Model

Model the use of the mathematical thinking map through class discussion with the thinking process modeled aloud as the thinking map is completed.

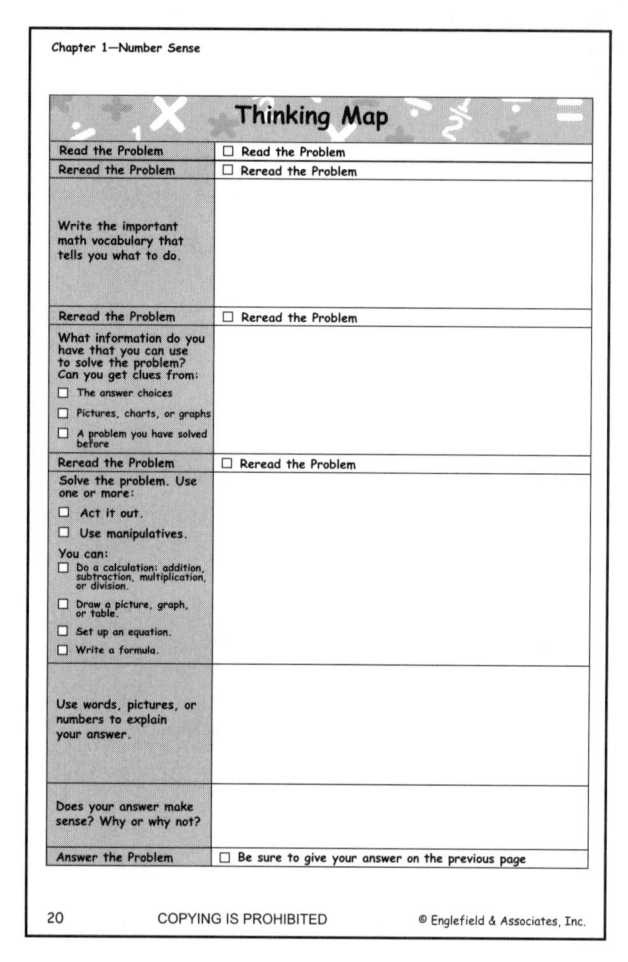

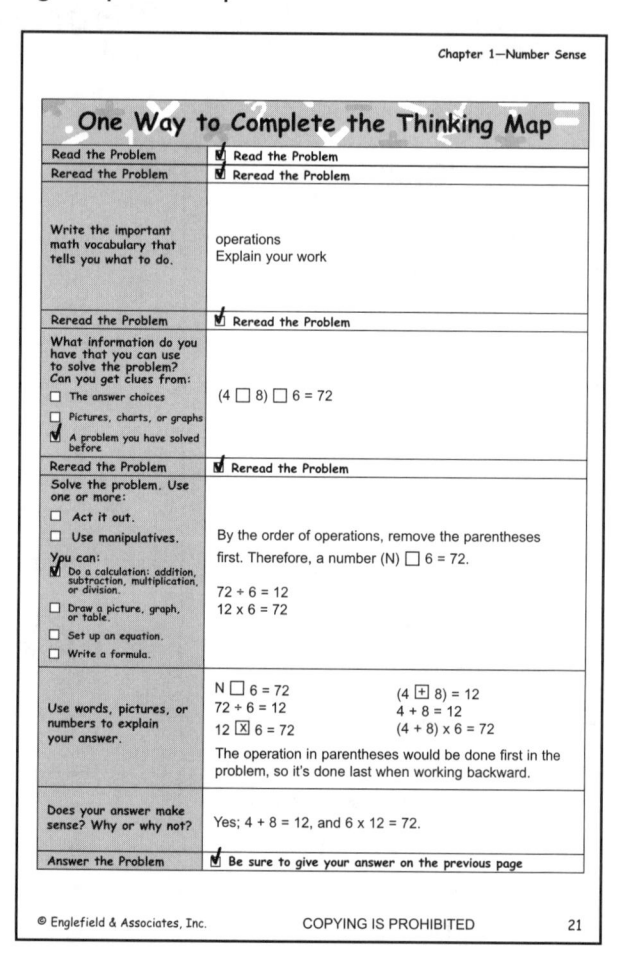

Thinking Map		One Way to Complete the Thinking Map	
Read the Problem	☐ Read the Problem	Read the Problem	☑ Read the Problem
Reread the Problem	☐ Reread the Problem	Reread the Problem	☑ Reread the Problem
Write the important math vocabulary that tells you what to do.		Write the important math vocabulary that tells you what to do.	operations Explain your work
Reread the Problem	☐ Reread the Problem	Reread the Problem	☑ Reread the Problem
What information do you have that you can use to solve the problem? Can you get clues from: ☐ The answer choices ☐ Pictures, charts, or graphs ☐ A problem you have solved before		What information do you have that you can use to solve the problem? Can you get clues from: ☐ The answer choices ☐ Pictures, charts, or graphs ☑ A problem you have solved before	(4 ☐ 8) ☐ 6 = 72
Reread the Problem	☐ Reread the Problem	Reread the Problem	☑ Reread the Problem
Solve the problem. Use one or more: ☐ Act it out. ☐ Use manipulatives. You can: ☐ Do a calculation: addition, subtraction, multiplication, or division. ☐ Draw a picture, graph, or table. ☐ Set up an equation. ☐ Write a formula.		Solve the problem. Use one or more: ☐ Act it out. ☐ Use manipulatives. You can: ☑ Do a calculation: addition, subtraction, multiplication, or division. ☐ Draw a picture, graph, or table. ☐ Set up an equation. ☐ Write a formula.	By the order of operations, remove the parentheses first. Therefore, a number (N) ☐ 6 = 72. 72 ÷ 6 = 12 12 × 6 = 72
Use words, pictures, or numbers to explain your answer.		Use words, pictures, or numbers to explain your answer.	N ☐ 6 = 72 (4 ⊞ 8) = 12 72 ÷ 6 = 12 4 + 8 = 12 12 ☒ 6 = 72 (4 + 8) × 6 = 72 The operation in parentheses would be done first in the problem, so it's done last when working backward.
Does your answer make sense? Why or why not?		Does your answer make sense? Why or why not?	Yes; 4 + 8 = 12, and 6 × 12 = 72.
Answer the Problem	☐ Be sure to give your answer on the previous page	Answer the Problem	☑ Be sure to give your answer on the previous page

3. Guided Practice

Provide guided practice by presenting a problem for students to complete that matches the strand, content, or process you want to teach.

4. Independent Practice

Provide independent practice so students are provided with multiple-choice, short-answer, or extended-response problems.

Chapter 1—Number Sense

Thinking Map

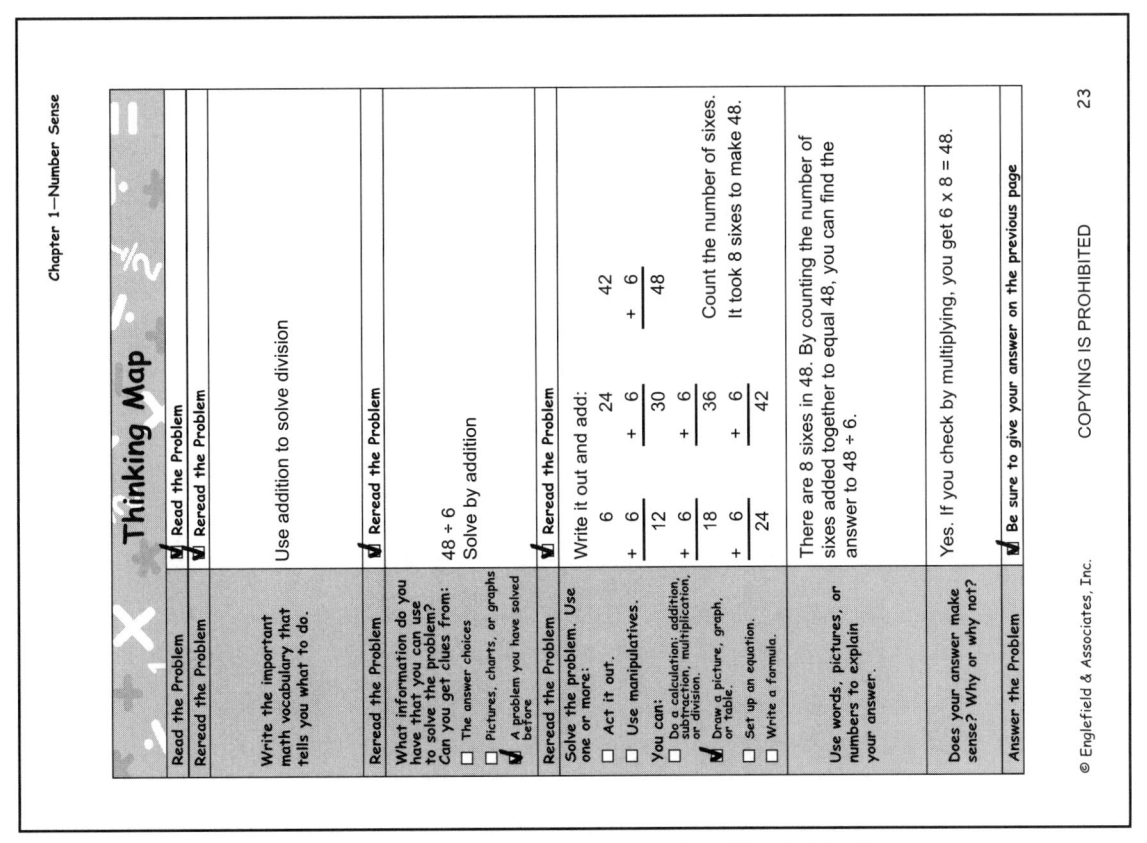

☑ Read the Problem ☑ Reread the Problem	Use addition to solve division
☑ Reread the Problem	$48 \div 6$ Solve by addition
☑ Reread the Problem	Write it out and add: $\begin{array}{r}6\\+\ 6\\\hline 12\end{array}$ $\begin{array}{r}24\\+\ 6\\\hline 30\end{array}$ $\begin{array}{r}42\\+\ 6\\\hline 48\end{array}$ $\begin{array}{r}6\\+\ 6\\\hline 18\end{array}$ $\begin{array}{r}30\\+\ 6\\\hline 36\end{array}$ $\begin{array}{r}18\\+\ 6\\\hline 24\end{array}$ $\begin{array}{r}36\\+\ 6\\\hline 42\end{array}$ Count the number of sixes. It took 8 sixes to make 48. There are 8 sixes in 48. By counting the number of sixes added together to equal 48, you can find the answer to $48 \div 6$.
	Yes. If you check by multiplying, you get $6 \times 8 = 48$.
☑ Answer the Problem	Be sure to give your answer on the previous page

In the Student Workbook, the Thinking Map is blank.
In the Parent/Teacher Edition, suggested ways for completing the Thinking Maps are presented.

Number Sense
Short-Answer Practice Problem 8

8. How can you use addition to solve $48 \div 6$?

1 point: The student recognizes that 6 is a factor of 48.

1 point: The student shows understanding of repeated addition: $6 + 6 + 6 + 6 + 6 + 6 + 6 + 6 = 48$.

Use the thinking map on the next page to solve the problem.
Write your answer in the box.

Thinking Map

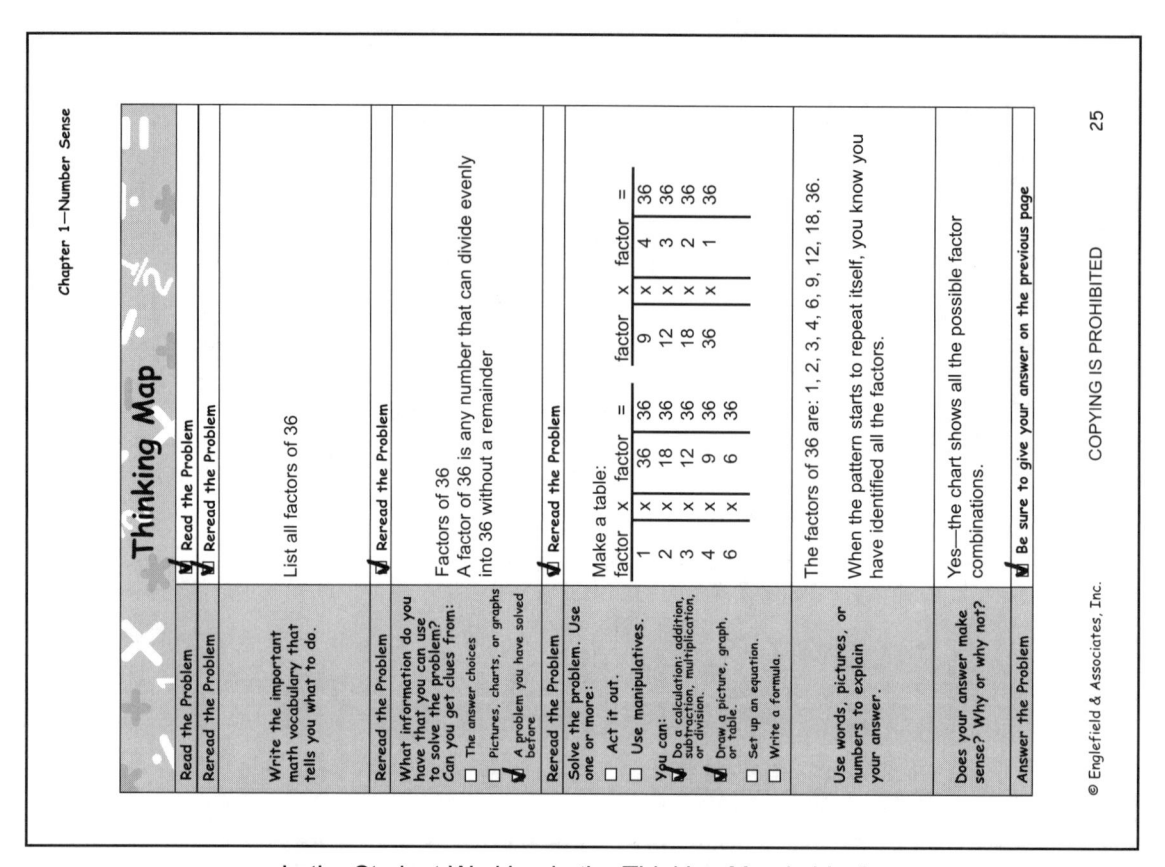

Read the Problem	☑ Read the Problem ☑ Reread the Problem List all factors of 36
Write the important math vocabulary that tells you what to do.	
Reread the Problem	☑ Reread the Problem Factors of 36 A factor of 36 is any number that can divide evenly into 36 without a remainder
What information do you have that you can use to solve the problem? Can you get clues from: ☐ The answer choices ☐ Pictures, charts, or graphs ☑ A problem you have solved before	
Reread the Problem	☑ Reread the Problem Make a table: factor × factor = 1 × 36 = 36 2 × 18 = 36 3 × 12 = 36 4 × 9 = 36 6 × 6 = 36 factor × factor = 9 × 4 = 36 12 × 3 = 36 18 × 2 = 36 36 × 1 = 36 The factors of 36 are: 1, 2, 3, 4, 6, 9, 12, 18, 36. When the pattern starts to repeat itself, you know you have identified all the factors.
Solve the problem. Use one or more: ☐ Act it out. ☐ Use manipulatives. You can: ☑ Do a calculation: addition, subtraction, multiplication, or division. ☑ Draw a picture, graph, or table. ☐ Set up an equation. ☐ Write a formula.	
Use words, pictures, or numbers to explain your answer.	Yes—the chart shows all the possible factor combinations.
Does your answer make sense? Why or why not?	
Answer the Problem	☑ Be sure to give your answer on the previous page

In the Student Workbook, the Thinking Map is blank.
In the Parent/Teacher Edition, suggested ways for completing the Thinking Maps are presented.

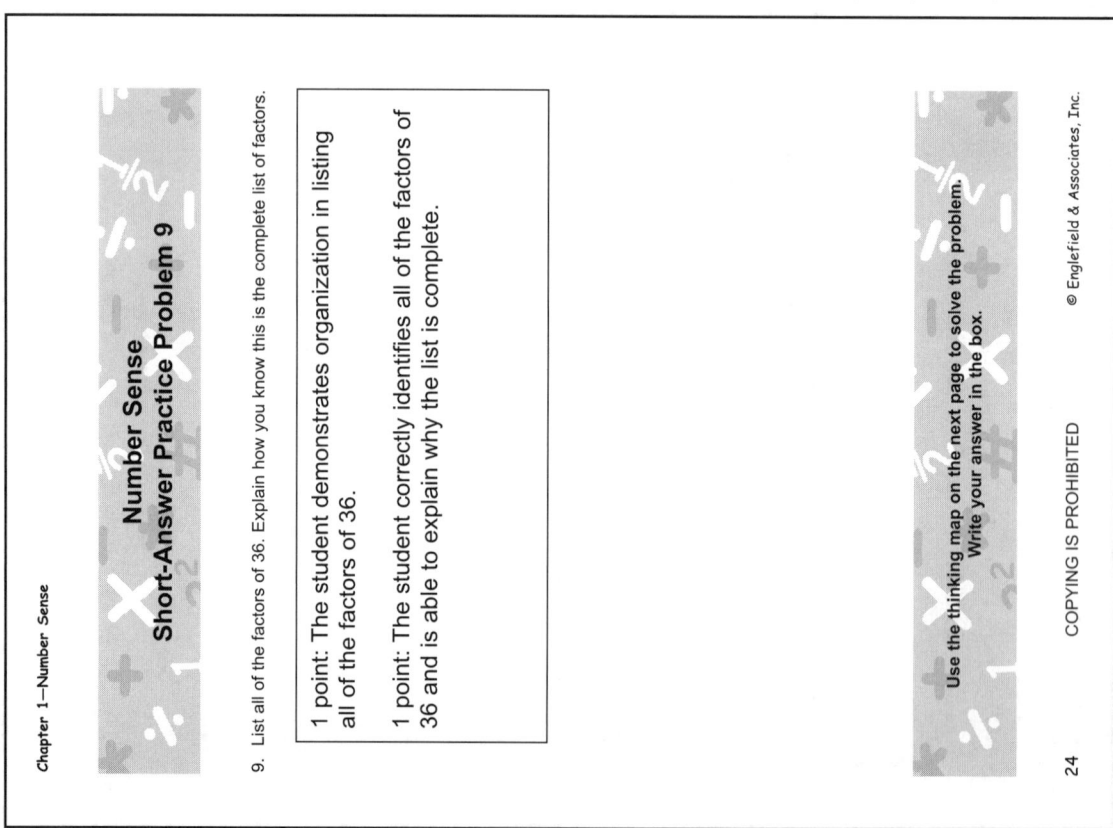

Number Sense
Short-Answer Practice Problem 9

9. List all of the factors of 36. Explain how you know this is the complete list of factors.

1 point: The student demonstrates organization in listing all of the factors of 36.

1 point: The student correctly identifies all of the factors of 36 and is able to explain why the list is complete.

Use the thinking map on the next page to solve the problem. Write your answer in the box.

Chapter 1—Number Sense

Model Problem 3: Number Sense Extended-Response

1. Prior Knowledge

Determine the student's prior knowledge about the strand, content of the problem, and/or mathematical process of the problem.

These questions can be used to help determine students' knowledge:

- When you make a purchase, how do you determine whether you received the correct change?

- What are some of the various ways cashiers determine a customer's change? (subtract, count up, use a calculator, etc.)

Additional prompts could be provided by using the following topics:

- a fast food restaurant
- a store without an electronic cash register
- a concession stand
- a fundraiser
- a yard sale

Chapter 1—Number Sense

Model Problem 3: Number Sense Extended-Response

3. Sarita bought an item that cost $13.76. She paid the cashier with a $20 bill. Draw a picture of the coins and bills Sarita should receive as change. What would the cashier say as he or she counts the change?

> 1 point: The student calculates the correct amount of change.
>
> 1 point: The student shows understanding of the bills and coins that would be used to give change.
>
> 1 point: The student draws the correct combination of bills and coins Sarita would receive as change.
>
> 1 point: The student correctly identifies how the cashier would count the change.

Use the thinking map on the next page to solve the problem.
Write your answer in the box.

Chapter 1—Number Sense

2. Model

Model the use of the mathematical thinking map through class discussion with the thinking process modeled aloud as the thinking map is completed.

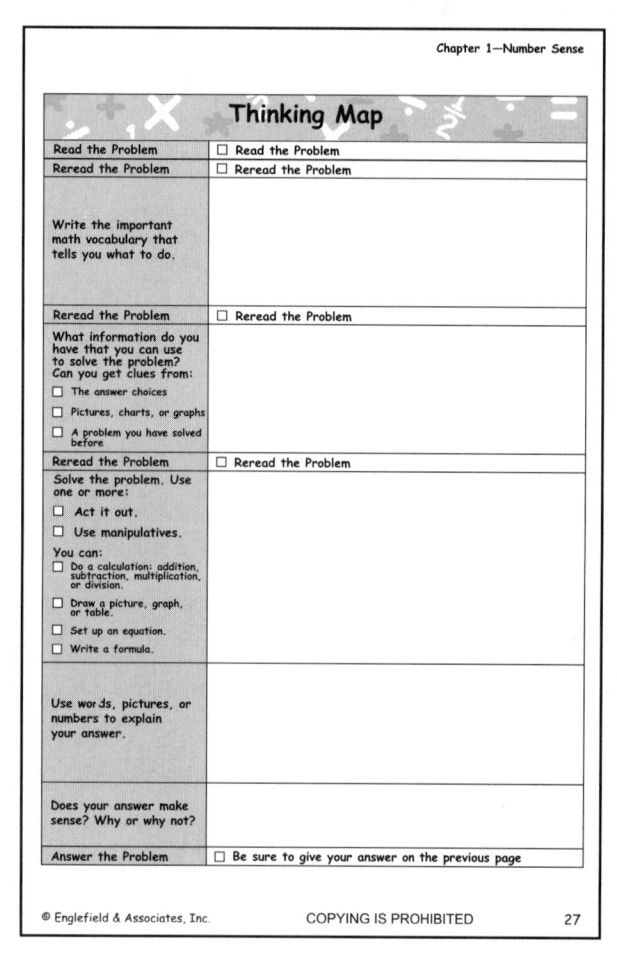

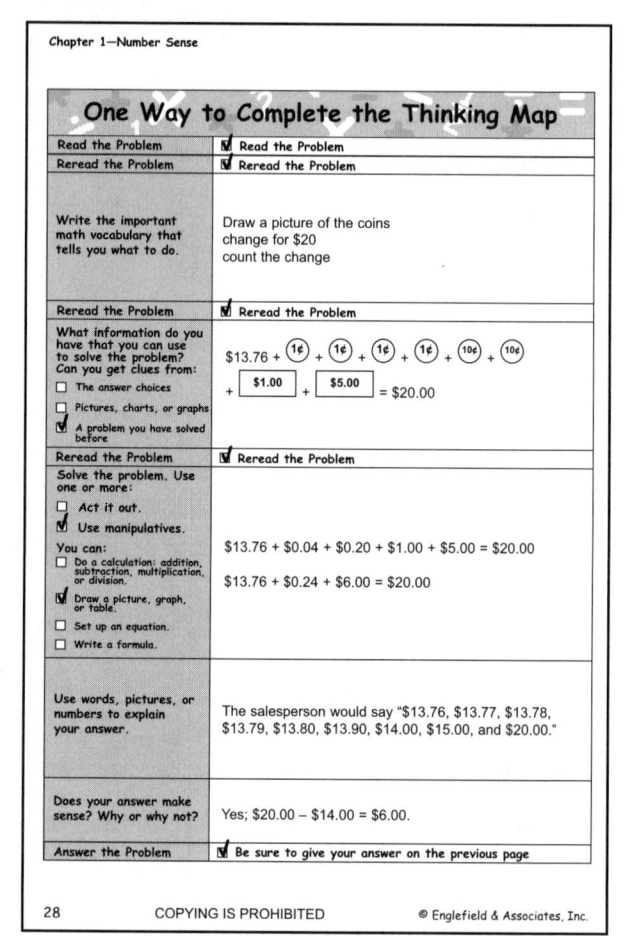

Thinking Map		One Way to Complete the Thinking Map	
Read the Problem	☐ Read the Problem	Read the Problem	☑ Read the Problem
Reread the Problem	☐ Reread the Problem	Reread the Problem	☑ Reread the Problem
Write the important math vocabulary that tells you what to do.		Write the important math vocabulary that tells you what to do.	Draw a picture of the coins change for $20 count the change
Reread the Problem	☐ Reread the Problem	Reread the Problem	☑ Reread the Problem
What information do you have that you can use to solve the problem? Can you get clues from: ☐ The answer choices ☐ Pictures, charts, or graphs ☐ A problem you have solved before		What information do you have that you can use to solve the problem? Can you get clues from: ☐ The answer choices ☐ Pictures, charts, or graphs ☑ A problem you have solved before	$13.76 + (1¢) + (1¢) + (1¢) + (1¢) + (10¢) + (10¢) + [$1.00] + [$5.00] = $20.00
Reread the Problem	☐ Reread the Problem	Reread the Problem	☑ Reread the Problem
Solve the problem. Use one or more: ☐ Act it out. ☐ Use manipulatives. You can: ☐ Do a calculation: addition, subtraction, multiplication, or division. ☐ Draw a picture, graph, or table. ☐ Set up an equation. ☐ Write a formula.		Solve the problem. Use one or more: ☐ Act it out. ☑ Use manipulatives. You can: ☐ Do a calculation: addition, subtraction, multiplication, or division. ☑ Draw a picture, graph, or table. ☐ Set up an equation. ☐ Write a formula.	$13.76 + $0.04 + $0.20 + $1.00 + $5.00 = $20.00 $13.76 + $0.24 + $6.00 = $20.00
Use words, pictures, or numbers to explain your answer.		Use words, pictures, or numbers to explain your answer.	The salesperson would say "$13.76, $13.77, $13.78, $13.79, $13.80, $13.90, $14.00, $15.00, and $20.00."
Does your answer make sense? Why or why not?		Does your answer make sense? Why or why not?	Yes; $20.00 − $14.00 = $6.00.
Answer the Problem	☐ Be sure to give your answer on the previous page	Answer the Problem	☑ Be sure to give your answer on the previous page

3. Guided Practice

Provide guided practice by presenting a problem for students to complete that matches the strand, content, or process you want to teach.

4. Independent Practice

Provide independent practice so students are provided with multiple-choice, short-answer, or extended-response problems.

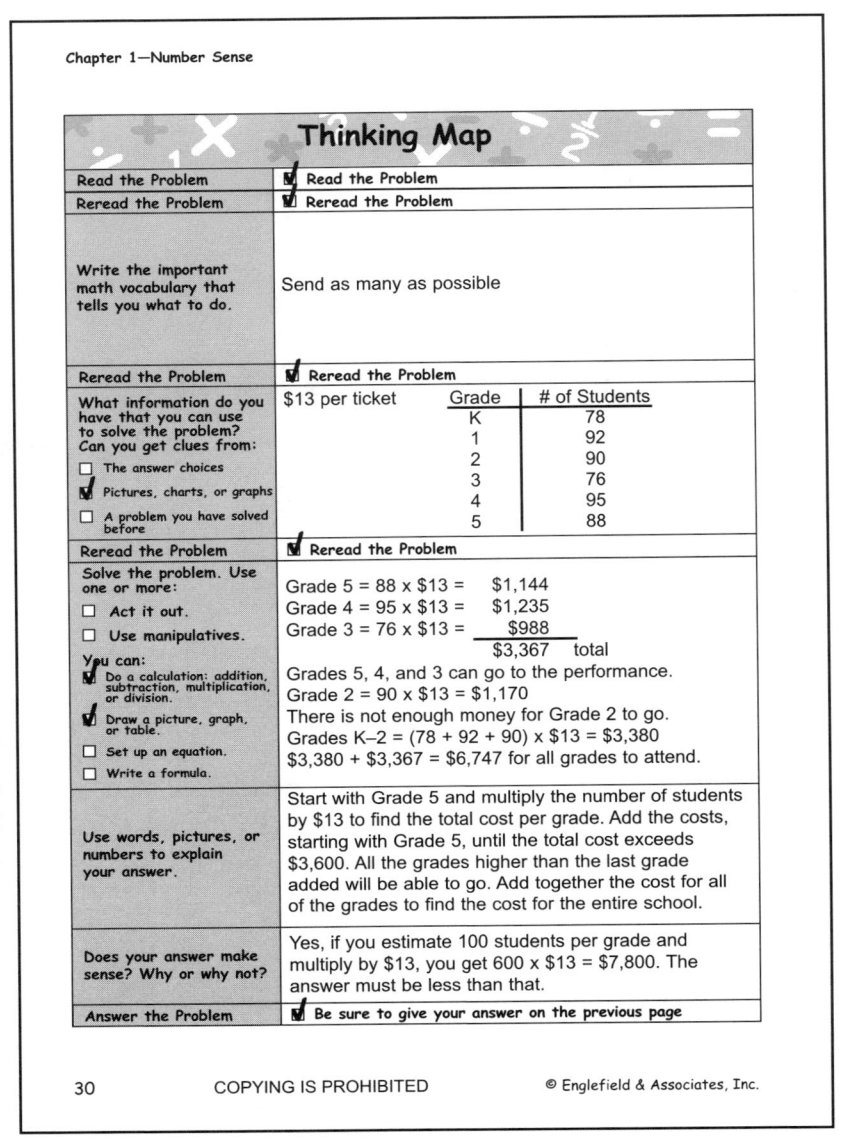

Chapter 1—Number Sense

Number Sense
Extended-Response Practice Problem 10

10. Tickets to The Lion King performance are $13.00 each. The principal of Southbrook Elementary School has $3,600.00 to purchase tickets. She decides to start with the fifth grade and send as many students as possible, then move to fourth grade, and so on. How many grades will be able to attend the performance? How much would it cost to send all of the students?

Grade	# of Students
Kindergarten	78
1	92
2	90
3	76
4	95
5	88

1 point: The student demonstrates organization and understanding of the problem-solving process.

1 point: The student computes the total cost of sending each grade.

1 point: The student determines that grades 3, 4, and 5 can attend the performance.

1 point: The student correctly computes the total cost for all students to attend.

Use the thinking map on the next page to solve the problem. Write your answer in the box.

In the Student Workbook, the Thinking Map is blank.
In the Parent/Teacher Edition, suggested ways for completing the Thinking Maps are presented.

Thinking Map

Read the Problem ☑

Reread the Problem ☑
Write the important math vocabulary that tells you what to do.
Send as many as possible

Reread the Problem ☑
What information do you have that you can use to solve the problem? Can you get clues from:
- ☐ The answer choices
- ☑ Pictures, charts, or graphs
- ☐ A problem you have solved before

$13 per ticket

Grade	# of Students
K	78
1	92
2	90
3	76
4	95
5	88

Reread the Problem ☑
Solve the problem. Use one or more:
- ☐ Act it out.
- ☐ Use manipulatives.

You can:
- ☑ Do calculation: addition, subtraction, multiplication, or division.
- ☑ Draw a picture, graph, or table.
- ☐ Set up an equation.
- ☐ Write a formula.

Grade 5 = 88 x $13 = $1,144
Grade 4 = 95 x $13 = $1,235
Grade 3 = 76 x $13 = $988
$3,367 total

Grades 5, 4, and 3 can go to the performance.
Grade 2 = 90 x $13 = $1,170
There is not enough money for Grade 2 to go.
Grades K–2 = (78 + 92 + 90) x $13 = $3,380
$3,380 + $3,367 = $6,747 for all grades to attend.

Use words, pictures, or numbers to explain your answer.
Start with Grade 5 and multiply the number of students by $13 to find the total cost per grade. Add the costs, starting with Grade 5, until the total cost exceeds $3,600. All the grades higher than the last grade added will be able to go. Add together the cost for all of the grades to find the cost for the entire school.

Does your answer make sense? Why or why not?
Yes, if you estimate 100 students per grade and multiply by $13, you get 600 x $13 = $7,800. The answer must be less than that.

Answer the Problem
Be sure to give your answer on the previous page

Literature Connections for Number Sense

Everybody Wins
Shelia Bruce
© 2001

What Comes in 2's, 3's, & 4's?
Suzanne Aker
© 1990

One Hundred Angry Ants
Elinor J. Pinczes
© 1999

Can You Count to a Googol?
Robert E. Wells
© 2000

Remainder of One
Elinor J. Pinczes
© 1995

Count Your Way Through Greece
Jim Haskins and Kathleen Benson
© 1996

Anno's Mysterious Multiplying Jar
Masaichiro Anno and Mitsumasa Anno
© 1999

The Hershey's Milk Chocolate Multiplication Book
Jerry Pallotta and Rob Bolster
© 2002

Eight Animals Bake a Cake
Susan Middleton Elya
© 2002

The Grapes of Math: Mind-Stretching Math Riddles
Greg Tang
© 2001

Six-Dinner Sid
Inga Moore
© 1993

Stacks of Trouble
Martha F. Brenner
© 2001

Amanda Bean's Amazing Dream: A Mathematical Story
Cindy Neuschwander, Marilyn Burns, et al
© 1998

I Know an Old Lady Who Swallowed a Fly
Colin Hawkins and Jacqui Hawkins
© 2004

Apple Fractions
Jerry Pallotta
© 2003

Funny & Fabulous Fraction Stories
Dan Greenberg
© 1999

The Phantom Tollbooth
Norton Juster
© 1961

Sluggers' Car Wash
Stuart J. Murphy
© 2002

How Much, How Many, How Far, How Heavy, How Long, How Tall is 1000?
Helen Nolan
© 1995

Midnight Math: Twelve Terrific Math Games
Peter Ledwon
© 2000

Chapter 2

Measurement

The purposes of this chapter include

1. Providing a definition of Measurement and what it means for fourth-grade students.
2. Offering teaching tips that reinforce Measurement concepts.
3. Offering multiple-choice, short-answer, and extended-response questions dealing with Measurement.
4. Providing ideas for connecting Measurement concepts and processes with other areas of the curriculum.

The following teaching tools are provided for extending students' thinking with Measurement:

- Model lesson for Measurement multiple-choice problems
- Seven Measurement multiple-choice practice problems
- Model lesson for Measurement short-answer problems
- Two Measurement short-answer practice problems
- Model lesson for Measurement extended-response problems
- One Measurement extended-response practice problem

What is Measurement?

Students are expected to measure an object and relate the number of measurement units to the size of the units used. Students are expected to describe perimeter, area, and volume. Students are expected to make simple unit conversions within a measurement system. Students are expected to identify appropriate units to measure perimeter, area, and volume. Students are expected to develop measurement strategies to write and solve multi-step measurement problems.

Chapter 2—Measurement

What does Measurement look like?

- Students compare the number of cups to fill a container to the number of gallons needed to fill the same container.
- Students describe perimeter, area, and volume.
- Students select appropriate units to measure perimeter, area, and volume.
- Students use string or links to find perimeter.
- Students use tiles or a grid to find area.
- Students use cubes to find volume.
- Students make simple unit conversions.
- Students solve multi-step measurement problems.

Vocabulary related to Measurement:

Angle: The amount of turning between two lines meeting at a common point.

Area: The amount of space a surface takes up; measured in square units. Some formulas for calculating area are: rectangle = length x width; triangle = 1/2 base x height; circle = π x radius squared, or πr^2 (π = 3.14).

Celsius: Also known as centigrade, a temperature scale developed by Anders Celsius in 1742 that uses the metric system to measure temperature (0° C = freezing point of water; 100° C = boiling point of water).

Coordinate plane: A plane determined by the intersection of two perpendicular number lines in which any point can be located.

Cube: A solid shape that has 6 square faces that are all equal in size, 8 vertices, and 12 equal edges.

Cubic unit of measurement: A standard unit for measuring volume or capacity, such as a cubic meter.

Degree: A unit for measuring the size of angles based on 360° in a complete circle.

Estimate: To make an approximate guess of a measurement using a familiar item such as a pencil, a paper clip, or another device.

Fahrenheit: A measurement scale that is used to measure temperature within the English (standard) measurement system. Gabriel Fahrenheit developed the scale in 1724. The U.S. weather reports on the news are based on this scale (32° F = freezing point of water; 212° F = boiling point of water).

Formula: A mathematical rule written using symbols, usually as an equation describing a certain relationship between quantities.

Gallon: A standard unit that is used for measuring the capacity of liquids, such as a gallon of milk. (1 gallon = 4 quarts = 8 pints = 16 cups)

Gram: A metric unit for measuring mass (weight). (1,000 grams = 1 kilogram)

Grid paper: Paper marked with evenly-spaced horizontal and vertical lines, used for graphing and drawing.

Length: The distance of an object from one end to the other end; length can be measured by a ruler.

Linear: Relating to a line; having one dimension in a straight direction; a linear function or relationship can be represented by a straight line graph.

Linear scale: A scale in which the divisions are evenly spaced, like a ruler.

Mass: The amount of matter in an object.

Metric system: A decimal system of measure that uses multiples of 10.

Metric units: Units of measurement based on the metric system. See chart below.

	Metric Units			Metric Units	
Length	cm	centimeter	Volume	cm^3	cubic centimeter
	m	meter		m^3	cubic meter
	dm	decimeter			
	km	kilometer	Mass	g	gram
				kg	kilogram
Area	cm^2	square centimeter			
	m^2	square meter			
	km^2	square kilometer	Temperature	°C	degrees Celsius
Capacity	mL	milliliter			
	L	liter			

Mile: A standard unit used to determine distances; equal to 5,280 feet.

Minute: A unit of time that is equal to 60 seconds. On an analog clock, this is equal to one rotation of the second hand.

Net: A flat shape that can be folded into a three-dimensional solid.

Ounce: A standard unit for measuring weight (16 oz = 1 lb).

Parallelogram: A quadrilateral with opposite sides that are parallel and of equal length and opposite angles that are equal.

Perimeter: The distance around the outside of a shape.

Plane: A flat surface.

Point: A position in space.

Polygon: A plane shape having three or more straight sides; e.g., triangle, square, diamond, octagon, rectangle.

Polyhedron: A three-dimensional shape with plane faces.

Pound: A standard unit for measuring weight (16 oz = 1 lb).

Chapter 2—Measurement

Prism: A solid, three-dimensional shape with two identical parallel bases; all other faces are rectangles. A prism takes its name from the shape of its base, e.g., triangular prism, rectangular prism, etc.

Quadrilateral: A polygon with four sides.

Rectangular prism: A three-dimensional shape that has six rectangular faces.

Second: A unit of time; 60 seconds equal one minute. Seconds are marked by one mark on the face of a clock or by one number on a digital clock that counts seconds.

Square units in measurement: Square units used to measure the area of a given surface, such as the surface of a floor being measured in square meters.

Standard system of measurement: A system of measurement most commonly used in the United States (also called customary system of measurement).

Standard units: Units of measurement based on the standard system (also called customary units). See chart below.

	Standard Units		Standard Units		Standard Units
Length	in inch yd yard ft foot mi mile	Capacity	c cup qt quart pt liter gal gallon	Volume	in^3 cubic inch ft^3 cubic foot yd^3 cubic yard
Area	in^2 square inch ft^2 square foot yd^2 square yard mi^2 square mile	Mass	oz ounce lb pound T ton	Temperature	°F degrees Fahrenheit

Temperature: The amount of heat that can be measured within an object.

Thermometer: An instrument used to measure the temperature of something—the outdoors, the liquid for an experiment, the body temperature of a person, etc.

Three-dimensional object: An object having the three dimensions of length, width, and height.

Time: A measurement from past to present to future; the measurement is performed with clocks, stopwatches, sundials, or other time instruments. (1 hour = 60 minutes; 1 minute = 60 seconds)

Volume (capacity): The amount of space that is occupied by a container or an object.

Weight: The measurement of gravity in the mass of an object.

Width: The breadth, or the distance, across an object from side to side.

Teaching Tips

- Provide students with measurement tools and allow them time to experiment with the tools.
- Provide manipulatives such as tiles, cubes, string, and links.
- Model the use of manipulatives to find perimeter, area, and volume.
- Provide students with authentic real-life measuring situations.
- Take students from the concrete (real objects) to the abstract (paper and pencil tasks) to help them understand unit conversion.

For Future Reference

Note different strategies students used to solve the problems.

Chapter 2—Measurement

Model Problem 4: Measurement Multiple-Choice

1. Prior Knowledge

Determine the student's prior knowledge about the strand, content of the problem, and/or mathematical process of the problem.

These questions can be used to help determine students' knowledge:

- When do you need to visualize a three-dimensional figure as flat (a net)?
- In what real-life situations would you need to visualize a three-dimensional figure as a net?

Additional prompts could be provided by using the following topics:

- sewing
- blueprints, diagrams, patterns
- maps, atlases
- art and advertising

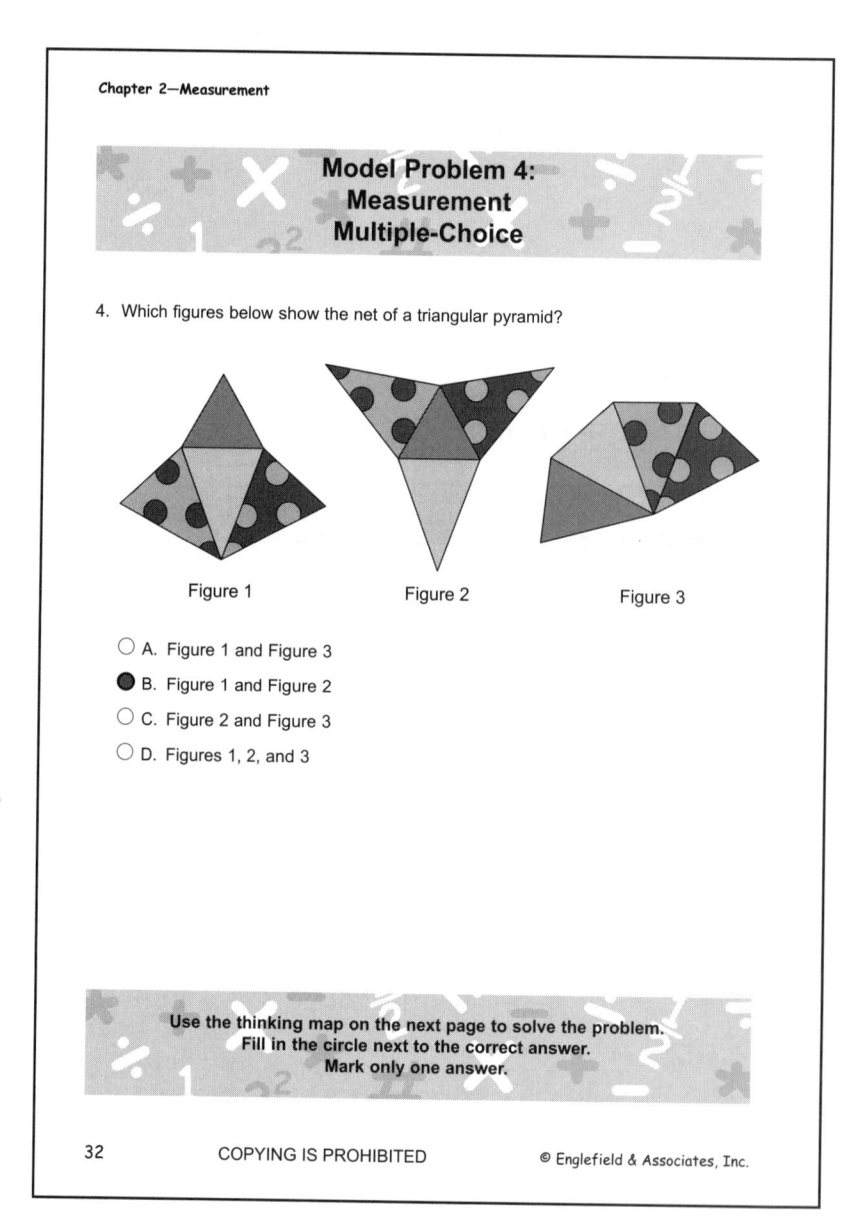

Chapter 2—Measurement

Model Problem 4: Measurement Multiple-Choice

4. Which figures below show the net of a triangular pyramid?

Figure 1   Figure 2   Figure 3

○ A. Figure 1 and Figure 3
● B. Figure 1 and Figure 2
○ C. Figure 2 and Figure 3
○ D. Figures 1, 2, and 3

Use the thinking map on the next page to solve the problem.
Fill in the circle next to the correct answer.
Mark only one answer.

Chapter 2—Measurement

2. Model

Model the use of the mathematical thinking map through class discussion with the thinking process modeled aloud as the thinking map is completed.

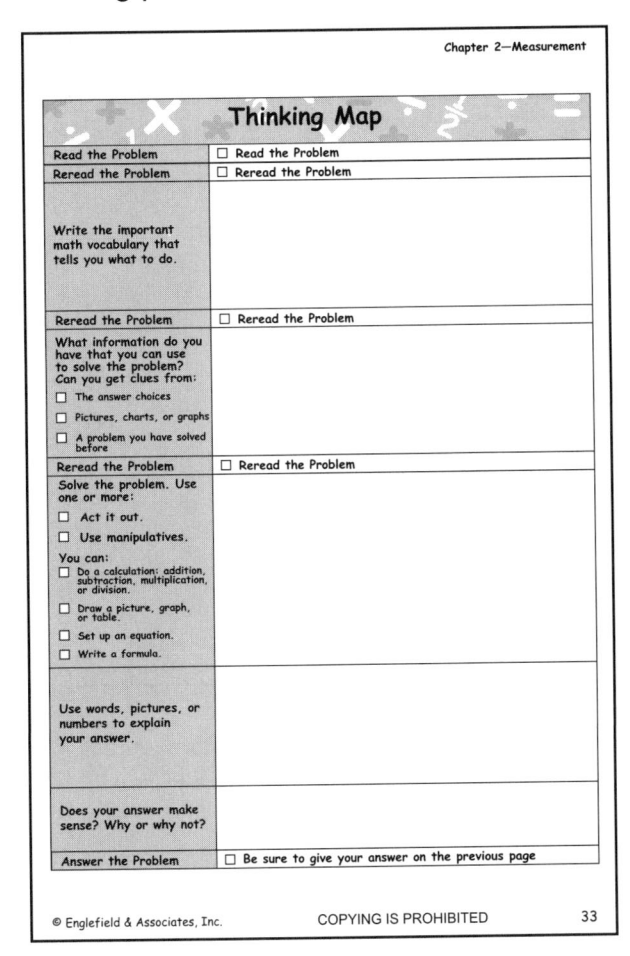

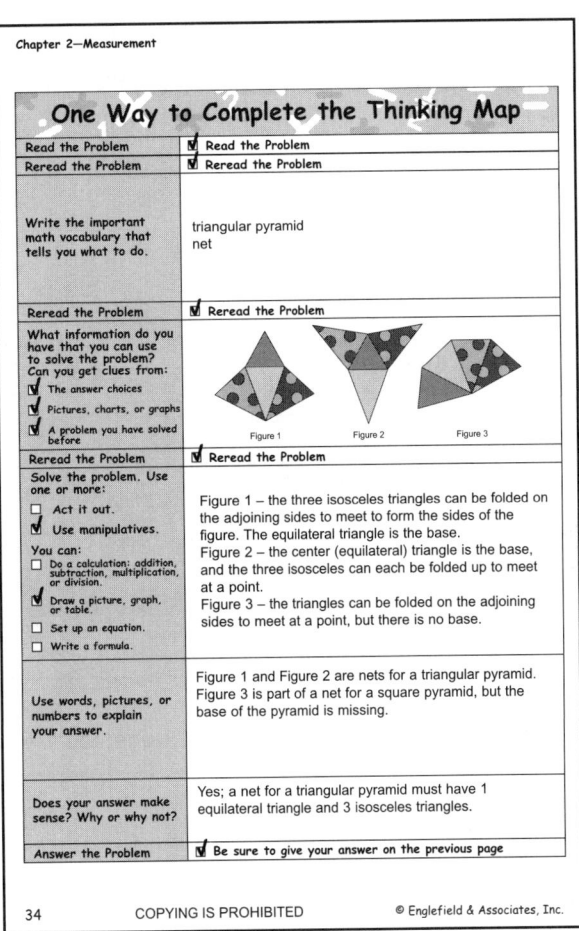

3. Guided Practice

Provide guided practice by presenting a problem for students to complete that matches the strand, content, or process you want to teach.

4. Independent Practice

Provide independent practice so students are provided with multiple-choice, short-answer, or extended-response problems.

© Englefield & Associates, Inc.   COPYING IS PROHIBITED   43

Chapter 2—Measurement

Thinking Map

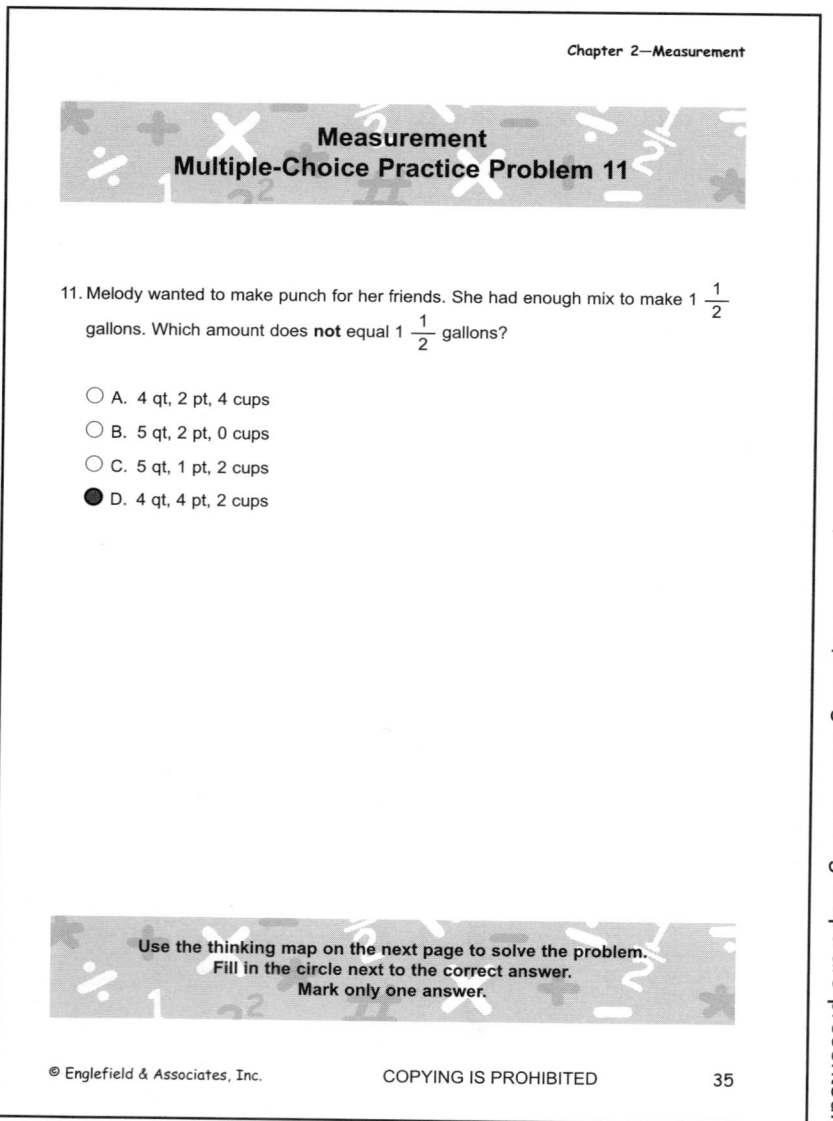

Read the Problem ☑ Read the Problem ☑ Reread the Problem	
Write the important math vocabulary that tells you what to do.	gallons pints quarts
Reread the Problem	☑ Reread the Problem
What information do you have that you can use to solve the problem? Can you get clues from:	$1\frac{1}{2}$ gallons 1 gallon = 4 quarts 2 pints = 1 quart 2 cups = 1 pint
☑ The answer choices ☑ Pictures, charts, or graphs ☐ A problem you have solved before	
Reread the Problem	☑ Reread the Problem
Solve the problem. Use one or more: ☐ Act it out. ☐ Use manipulatives. You can: ☑ Do a calculation: addition, subtraction, multiplication, or division. ☑ Draw a picture, graph, or table. ☐ Set up an equation. ☐ Write a formula.	4 qt + 2 pt + 4 cups = 1 gal + 1 qt + 1 qt + 1 1/2 gal 5 qt + 2 pt + 0 cups = 1 gal + 1 qt + 1 qt = 1 1/2 gal 5 qt + 1 pt + 2 cups = 1 gal + 1 qt + 1 pt + 1 pt = 1 1/2 gal 4 qt + 4 pt + 2 cups = 1 gal + 2 qt + 1 pt = 1 1/2 gal + 1 pt 1 1/2 gal + 1 pt ≠ 1 1/2 gal 4 qt + 4 pt + 2 cups ≠ 1 1/2 gal
Use words, pictures, or numbers to explain your answer.	
Does your answer make sense? Why or why not?	Yes; 1 gallon + 2 quarts = 1 1/2 gallons. Answer D adds up to more than 1 1/2 gallons.
Answer the Problem	☑ Be sure to give your answer on the previous page

36

In the Student Workbook, the Thinking Map is blank.
In the Parent/Teacher Edition, suggested ways for completing the Thinking Maps are presented.

Measurement
Multiple-Choice Practice Problem 11

11. Melody wanted to make punch for her friends. She had enough mix to make $1\frac{1}{2}$ gallons. Which amount does **not** equal $1\frac{1}{2}$ gallons?

○ A. 4 qt, 2 pt, 4 cups
○ B. 5 qt, 2 pt, 0 cups
○ C. 5 qt, 1 pt, 2 cups
● D. 4 qt, 4 pt, 2 cups

Use the thinking map on the next page to solve the problem.
Fill in the circle next to the correct answer.
Mark only one answer.

35

Measurement
Multiple-Choice Practice Problem 12

12. Sydney had a sleepover for her birthday. Her friends arrived at 7:00 p.m., and Sydney's mother told the girls that midnight was lights-out time. Sydney and her mother had planned seven activities for the party. The chart below shows the activities and the amount of time taken for each of the first five activities.

ACTIVITY	TIME
Introductions	30 min
Cake & Ice Cream	35 min
Game 1	22 min
Game 2	34 min
Movie	1 hour 57 min
Bedtime Snack	?
Story	?

After the first five activities, **about** how much total time was left for bedtime snack and story before lights out?

○ A. 2 hours
● B. 1 hour
○ C. $1\frac{1}{2}$ hours
○ D. $\frac{1}{2}$ hour

Use the thinking map on the next page to solve the problem.
Fill in the circle next to the correct answer.
Mark only one answer.

Thinking Map

☑ **Read the Problem**	
☑ **Reread the Problem**	7:00 p.m. midnight hours minutes
Write the important math vocabulary that tells you what to do.	
☑ **Reread the Problem**	Start at 7:00 p.m. Lights out at midnight 30 min., 25 min., 22 min., 34 min., 1 hour 57 min. About how much time was left?
What information do you use to solve the problem? Can you get clues from: ☐ The answer choices ☐ Pictures, charts, or graphs ☐ A problem you have solved before	
☑ **Reread the Problem**	30 min. + 25 min. + 22 min. + 34 min. = 121 min. 121 min. is about 2 hours
Solve the problem. Use one or more: ☐ Act it out. ☐ Use manipulatives. You can: ☑ Do a calculation: addition, subtraction, multiplication, or division. ☐ Draw a picture, graph, or table. ☐ Set up an equation. ☐ Write a formula.	1 hour 57 min. is about 2 hours 2 hours + 2 hours = 4 hours; 7:00 + 4 hours = 11:00 p.m. 1 hour is left for snack time and story time.
Use words, pictures, or numbers to explain your answer.	Adding up the times, 30 min. + 25 min. + 22 min. + 34 min. + 1 hour 57 min., shows that the activities take up about 4 hours and will last until about 11:00 p.m. That leaves 1 hour before midnight for the last two activities.
Does your answer make sense? Why or why not?	Yes—from 7:00 p.m. to midnight is 5 hours. If the activities listed take 4 hours, that leaves 1 hour.
☑ **Answer the Problem**	Be sure to give your answer on the previous page

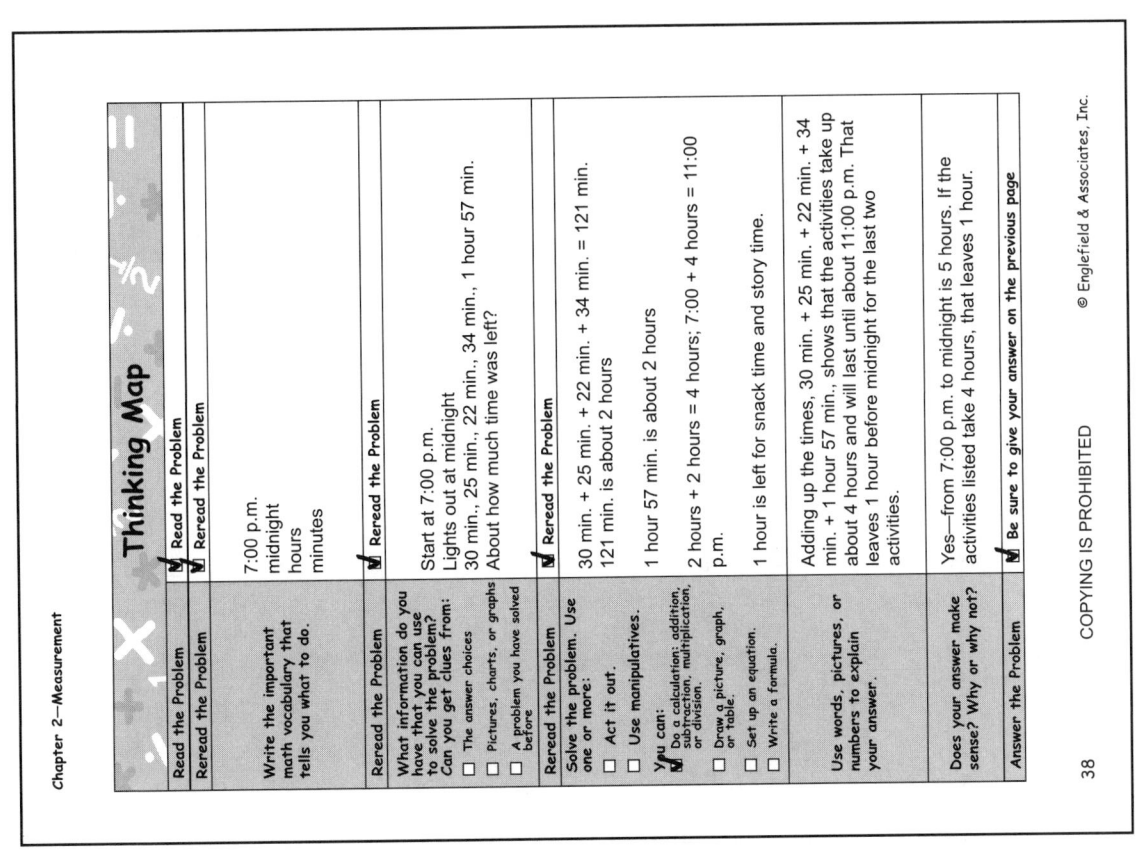

In the Student Workbook, the Thinking Map is blank.
In the Parent/Teacher Edition, suggested ways for completing the Thinking Maps are presented.

Chapter 2—Measurement

Thinking Map

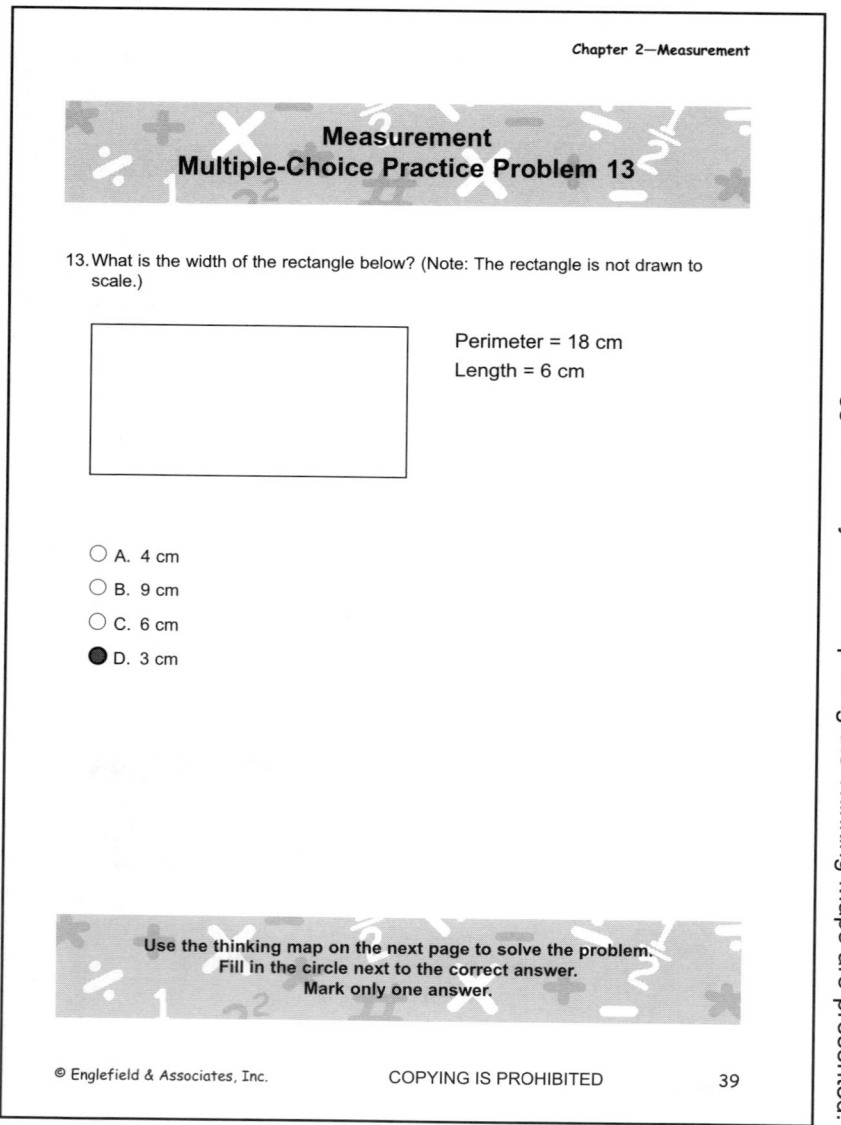

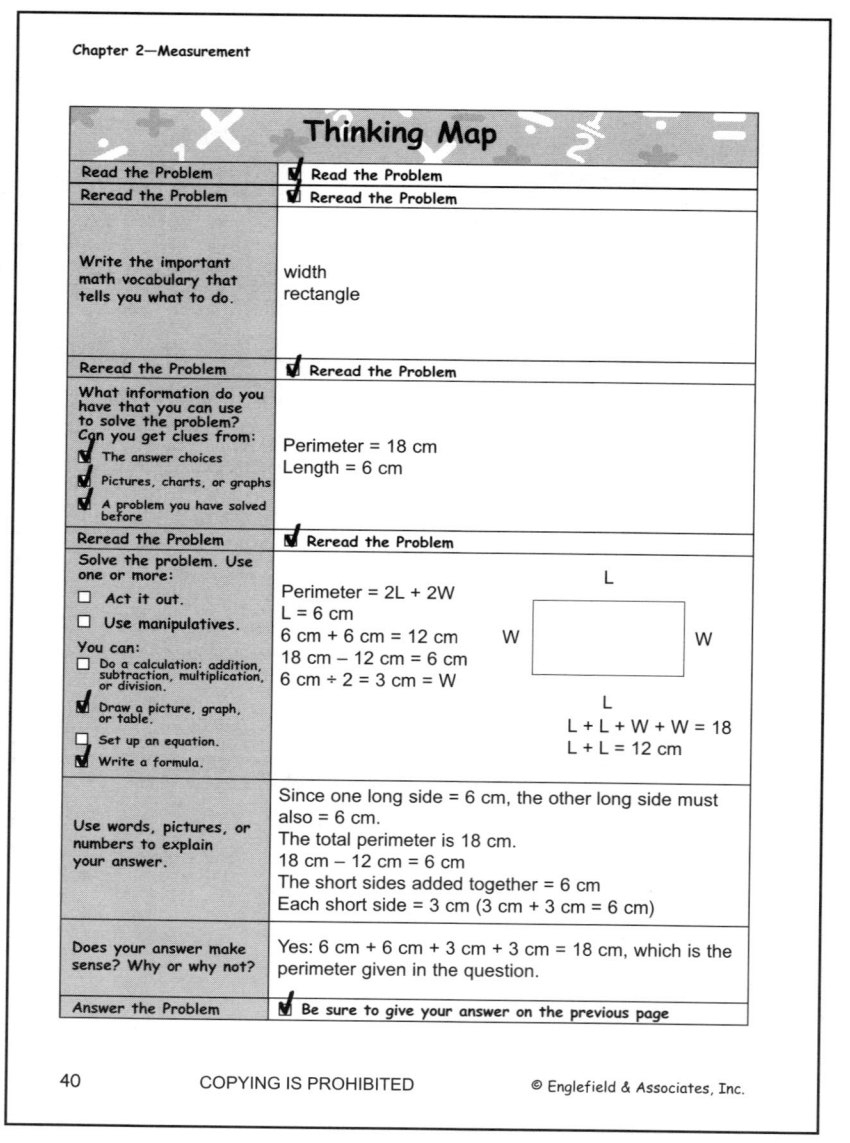

Read the Problem	☑ Read the Problem
	☑ Reread the Problem
Write the important math vocabulary that tells you what to do.	width rectangle
Reread the Problem	☑ Reread the Problem
What information do you have that you can use to solve the problem? Can you get clues from: ☐ The answer choices ☑ Pictures, charts, or graphs ☑ A problem you have solved before	Perimeter = 18 cm Length = 6 cm
Reread the Problem	☑ Reread the Problem
Solve the problem. Use one or more: ☐ Act it out. ☐ Use manipulatives. You can: ☐ Do a calculation: addition, subtraction, multiplication, or division. ☑ Draw a picture, graph, or table. ☑ Set up an equation. ☑ Write a formula.	Perimeter = 2L + 2W L = 6 cm 6 cm + 6 cm = 12 cm 18 cm − 12 cm = 6 cm 6 cm ÷ 2 = 3 cm = W W [rectangle] L L + L + W + W = 18 L + L = 12 cm
Use words, pictures, or numbers to explain your answer.	Since one long side = 6 cm, the other long side must also = 6 cm. The total perimeter is 18 cm. 18 cm − 12 cm = 6 cm The short sides added together = 6 cm Each short side = 3 cm (3 cm + 3 cm = 6 cm)
Does your answer make sense? Why or why not?	Yes: 6 cm + 6 cm + 3 cm + 3 cm = 18 cm, which is the perimeter given in the question.
Answer the Problem	☑ Be sure to give your answer on the previous page

In the Student Workbook, the Thinking Map is blank.
In the Parent/Teacher Edition, suggested ways for completing the Thinking Maps are presented.

Measurement
Multiple-Choice Practice Problem 13

13. What is the width of the rectangle below? (Note: The rectangle is not drawn to scale.)

Perimeter = 18 cm
Length = 6 cm

○ A. 4 cm
○ B. 9 cm
○ C. 6 cm
● D. 3 cm

Use the thinking map on the next page to solve the problem.
Fill in the circle next to the correct answer.
Mark only one answer.

Thinking Map

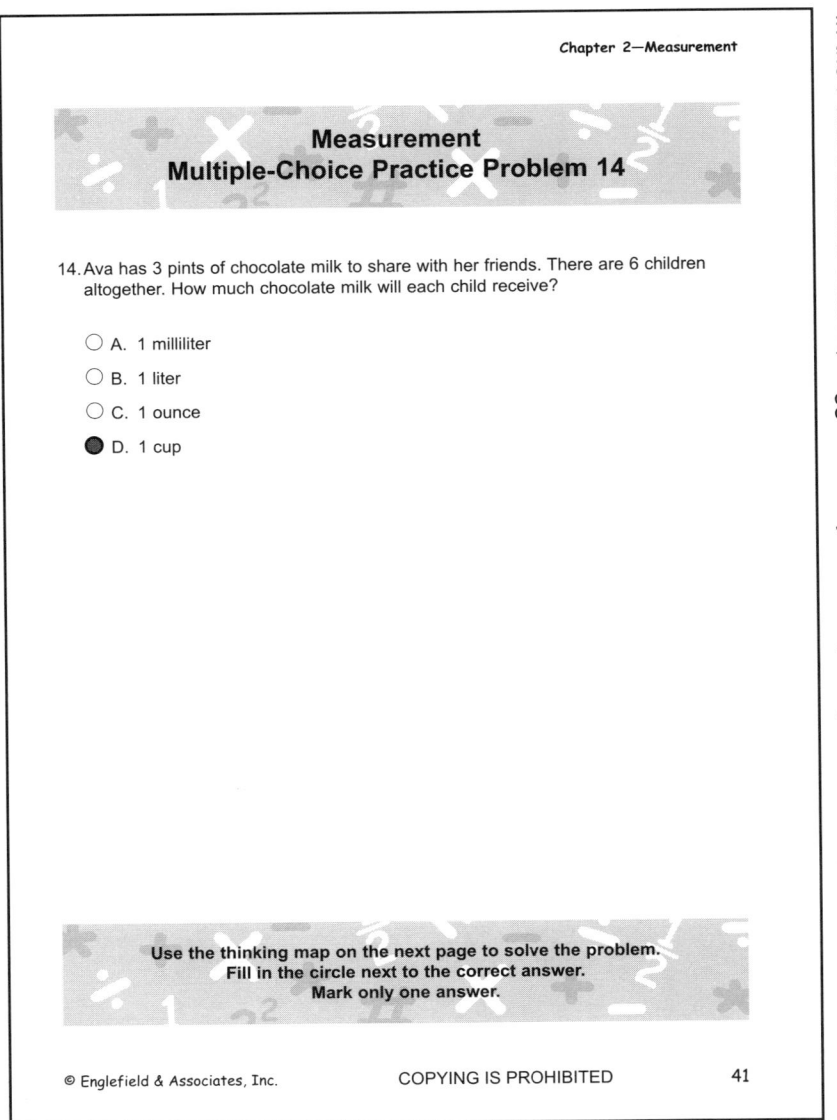

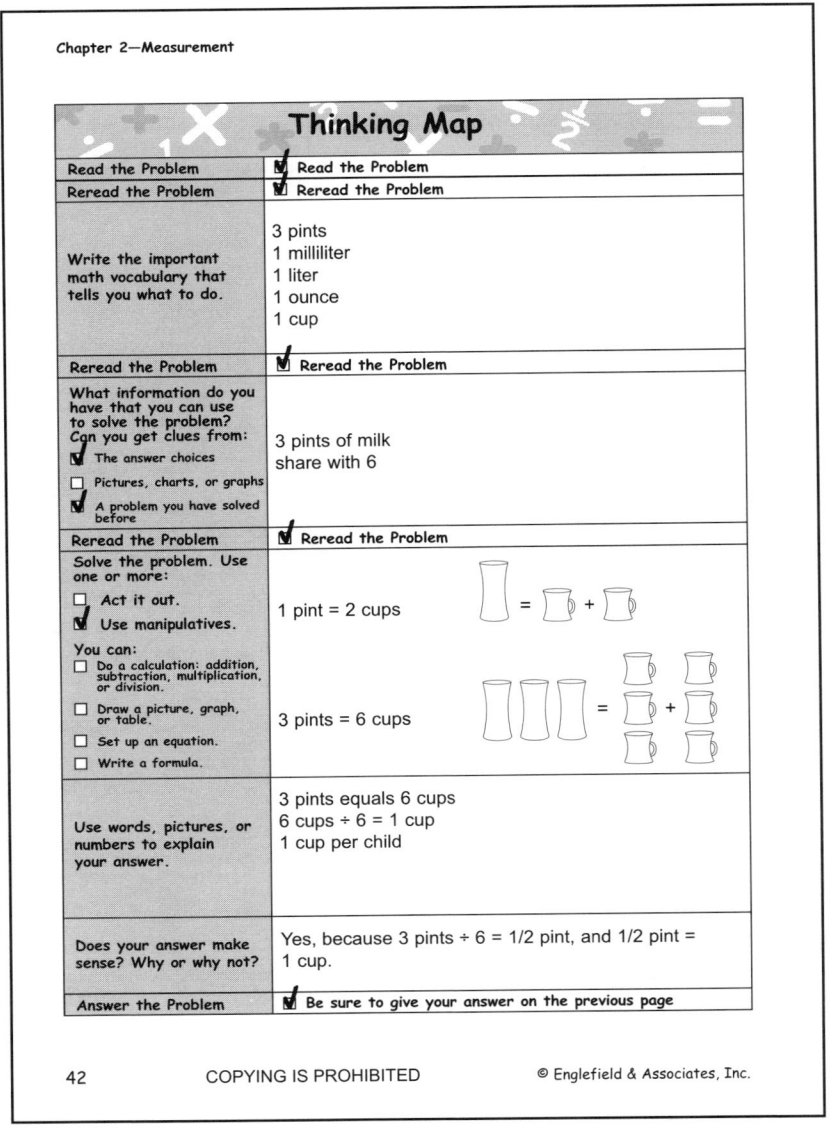

☑ Read the Problem	
☑ Reread the Problem	
Write the important math vocabulary that tells you what to do.	3 pints 1 milliliter 1 liter 1 ounce 1 cup
☑ Reread the Problem	
What information do you have that you can use to solve the problem? Can you get clues from: ☐ The answer choices ☐ Pictures, charts, or graphs ☑ A problem you have solved before	3 pints of milk share with 6
☑ Reread the Problem	
Solve the problem. Use one or more: ☐ Act it out. ☑ Use manipulatives. You can: ☐ Do a calculation: addition, subtraction, multiplication, or division. ☐ Draw a picture, graph, or table. ☐ Set up an equation. ☐ Write a formula.	1 pint = 2 cups 3 pints = 6 cups 3 pints equals 6 cups 6 cups ÷ 6 = 1 cup 1 cup per child
Use words, pictures, or numbers to explain your answer.	
Does your answer make sense? Why or why not?	Yes, because 3 pints ÷ 6 = 1/2 pint, and 1/2 pint = 1 cup.
☑ Answer the Problem	Be sure to give your answer on the previous page

In the Student Workbook, the Thinking Map is blank.
In the Parent/Teacher Edition, suggested ways for completing the Thinking Maps are presented.

Measurement
Multiple-Choice Practice Problem 14

14. Ava has 3 pints of chocolate milk to share with her friends. There are 6 children altogether. How much chocolate milk will each child receive?

○ A. 1 milliliter
○ B. 1 liter
○ C. 1 ounce
● D. 1 cup

Use the thinking map on the next page to solve the problem.
Fill in the circle next to the correct answer.
Mark only one answer.

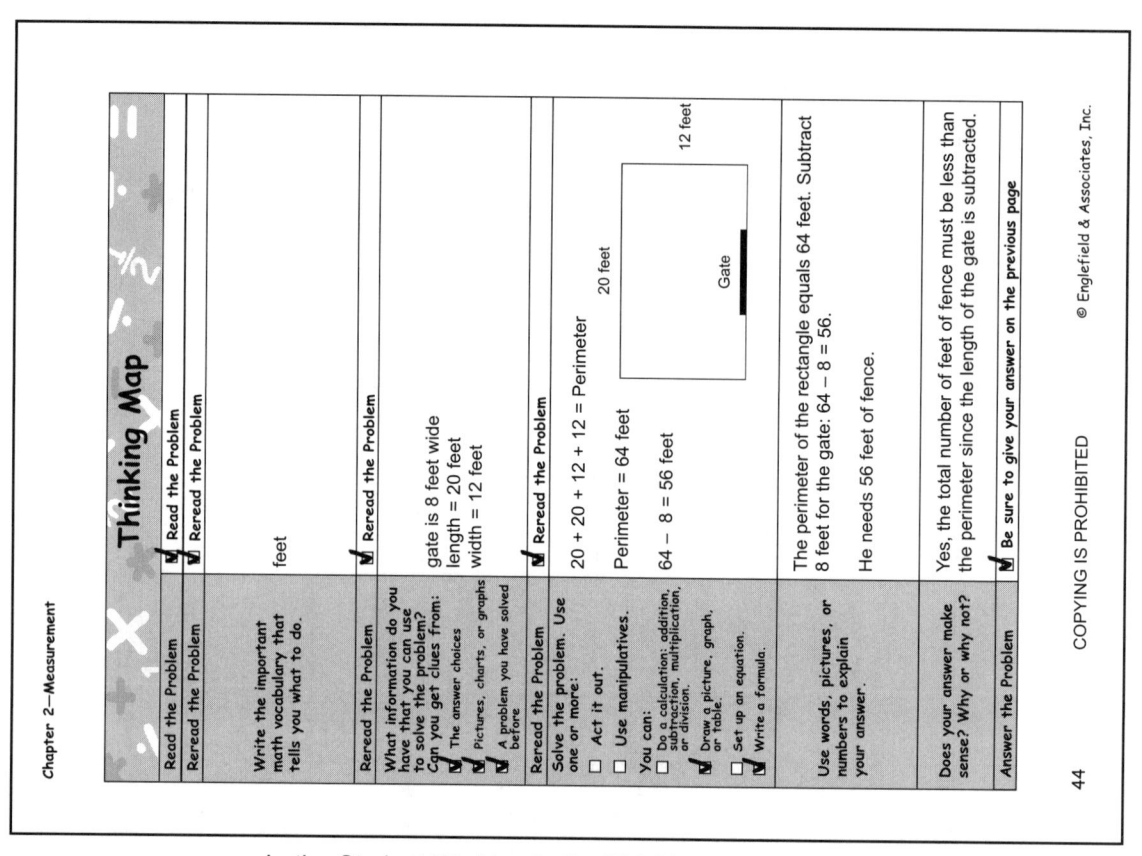

Chapter 2—Measurement

Thinking Map

Read the Problem	☑ Read the Problem ☑ Reread the Problem
Reread the Problem	
Write the important math vocabulary that tells you what to do.	feet
Reread the Problem	☑ Reread the Problem
What information do you have that you can use to solve the problem? Can you get clues from: ☐ The answer choices ☐ Pictures, charts, or graphs ☑ A problem you have solved before	gate is 8 feet wide length = 20 feet width = 12 feet
Reread the Problem	☑ Reread the Problem
Solve the problem. Use one or more: ☐ Act it out. ☐ Use manipulatives. You can: ☐ Do a calculation: addition, subtraction, multiplication, or division. ☑ Draw a picture, graph, or table. ☐ Set up an equation. ☑ Write a formula.	20 + 20 + 12 + 12 = Perimeter Perimeter = 64 feet 64 − 8 = 56 feet [Rectangle: 20 feet × 12 feet, with Gate]
Use words, pictures, or numbers to explain your answer.	The perimeter of the rectangle equals 64 feet. Subtract 8 feet for the gate: 64 − 8 = 56. He needs 56 feet of fence.
Does your answer make sense? Why or why not?	Yes, the total number of feet of fence must be less than the perimeter since the length of the gate is subtracted.
Answer the Problem	☑ Be sure to give your answer on the previous page

44 COPYING IS PROHIBITED © Englefield & Associates, Inc.

In the Student Workbook, the Thinking Map is blank.
In the Parent/Teacher Edition, suggested ways for completing the Thinking Maps are presented.

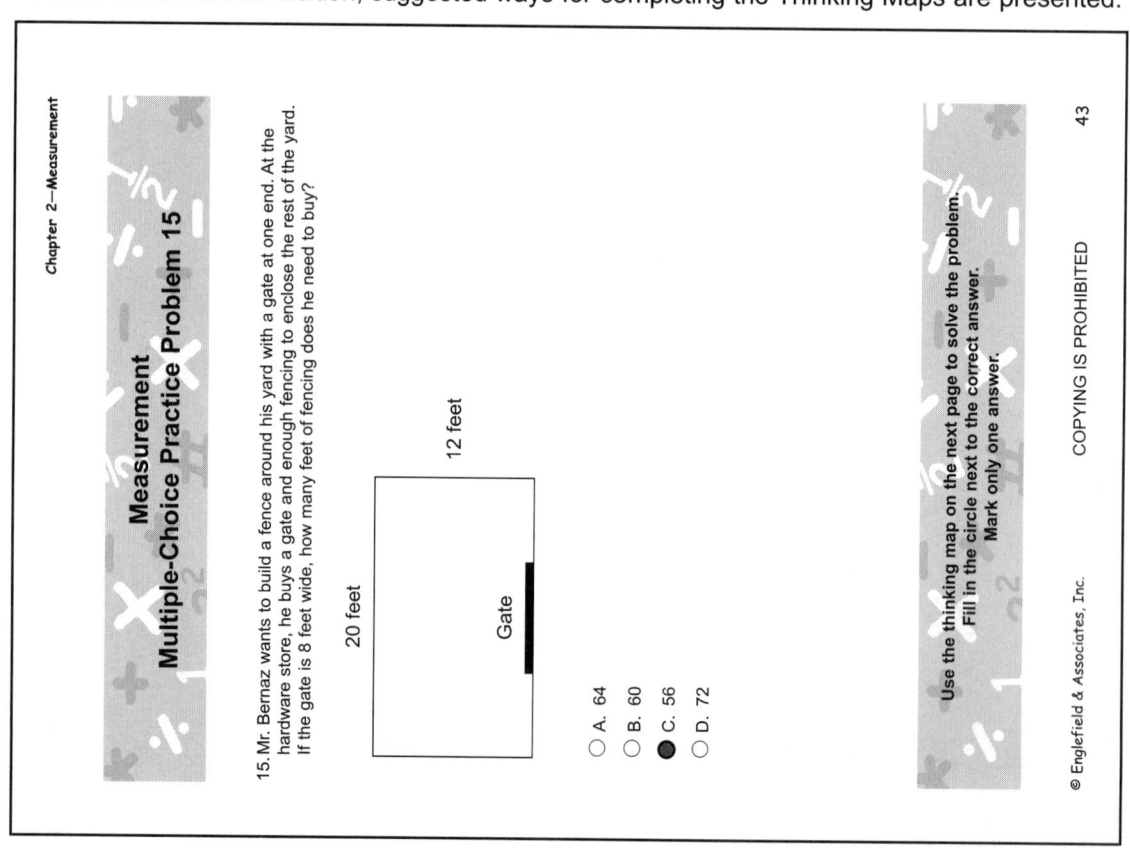

Chapter 2—Measurement

Measurement
Multiple-Choice Practice Problem 15

15. Mr. Bernaz wants to build a fence around his yard with a gate at one end. At the hardware store, he buys a gate and enough fencing to enclose the rest of the yard. If the gate is 8 feet wide, how many feet of fencing does he need to buy?

[Rectangle: 20 feet × 12 feet, with Gate]

○ A. 64
○ B. 60
● C. 56
○ D. 72

Use the thinking map on the next page to solve the problem.
Fill in the circle next to the correct answer.
Mark only one answer.

© Englefield & Associates, Inc.   COPYING IS PROHIBITED   43

Chapter 2—Measurement

Thinking Map

Chapter 2—Measurement

☑ Read the Problem	
☑ Reread the Problem	
Write the important math vocabulary that tells you what to do.	divide into equal areas length yards feet
☑ Reread the Problem	
What information do you have that you can use to solve the problem? Can you get clues from: ☑ The answer choices ☐ Pictures, charts, or graphs ☐ A problem you have solved before	Measurements are given in yards and feet. Must divide into two equal parts. Length of driveway is 11 yards.
☑ Reread the Problem	
Solve the problem. Use one or more: ☐ Act it out. ☐ Use manipulatives. You can: ☐ Do a calculation: addition, subtraction, multiplication, or division. ☑ Draw a picture, graph, or table. ☐ Set up an equation. ☐ Write a formula.	driveway = 11 yards | 5 1/2 yards | 5 1/2 yards | Answer is in feet—change yards to feet driveway = 11 yards x 3 feet = 33 feet | 16 1/2 feet | 16 1/2 feet | The driveway is 33 feet long. 33 ÷ 2 = 16.5 or 16 1/2. Each section of the driveway is 16 1/2 feet.
Use words, pictures, or numbers to explain your answer.	
Does your answer make sense? Why or why not?	Yes. Half of 11 yards = 5 1/2 yards. 5 yards = 15 feet, so 5 1/2 yards should be slightly longer than 15 feet.
Answer the Problem	☑ Be sure to give your answer on the previous page

46 COPYING IS PROHIBITED © Englefield & Associates, Inc.

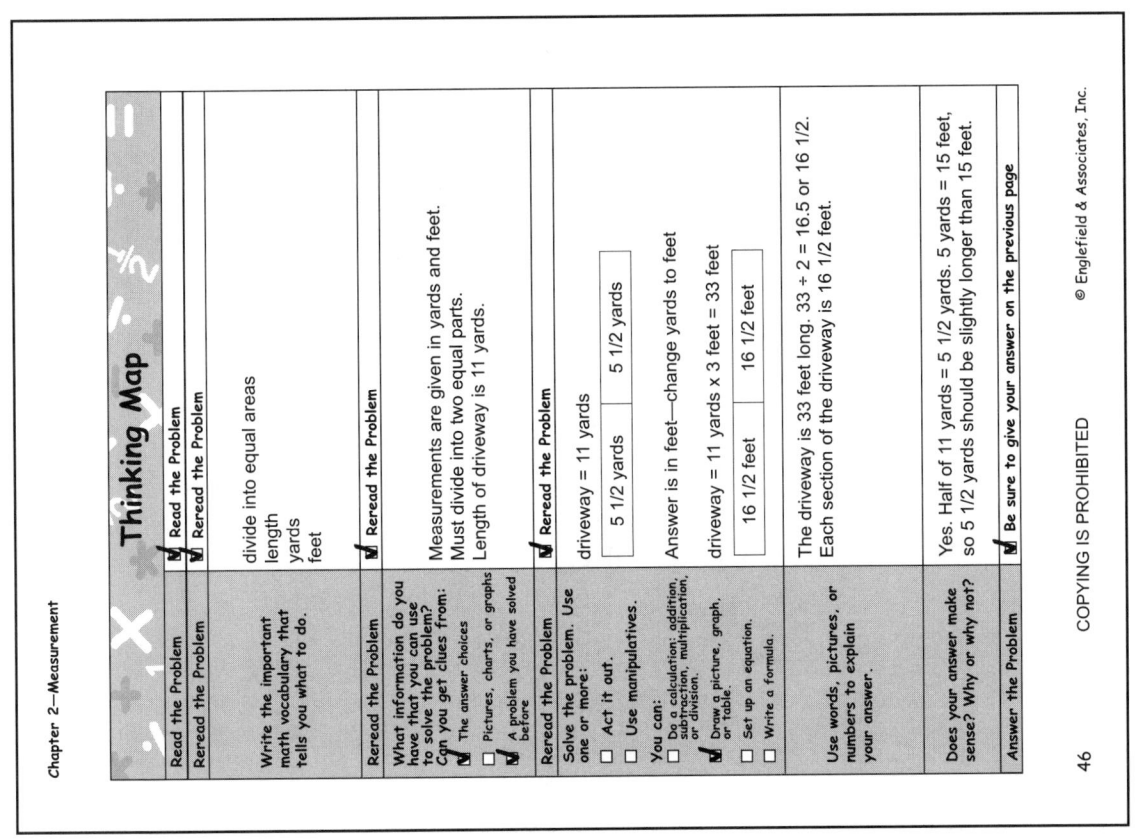

In the Student Workbook, the Thinking Map is blank.
In the Parent/Teacher Edition, suggested ways for completing the Thinking Maps are presented.

Chapter 2—Measurement

Measurement
Multiple-Choice Practice Problem 16

16. Maggie wants to divide her driveway into two areas of equal size for playing hopscotch and jumping rope. The length of her driveway is 11 yards. What will be the length, in feet, of each section?

○ A. 15 feet

○ B. 16 feet

● C. $16 \frac{1}{2}$ feet

○ D. $15 \frac{1}{2}$ feet

Use the thinking map on the next page to solve the problem.
Fill in the circle next to the correct answer.
Mark only one answer.

© Englefield & Associates, Inc. COPYING IS PROHIBITED 45

Chapter 2—Measurement

Thinking Map

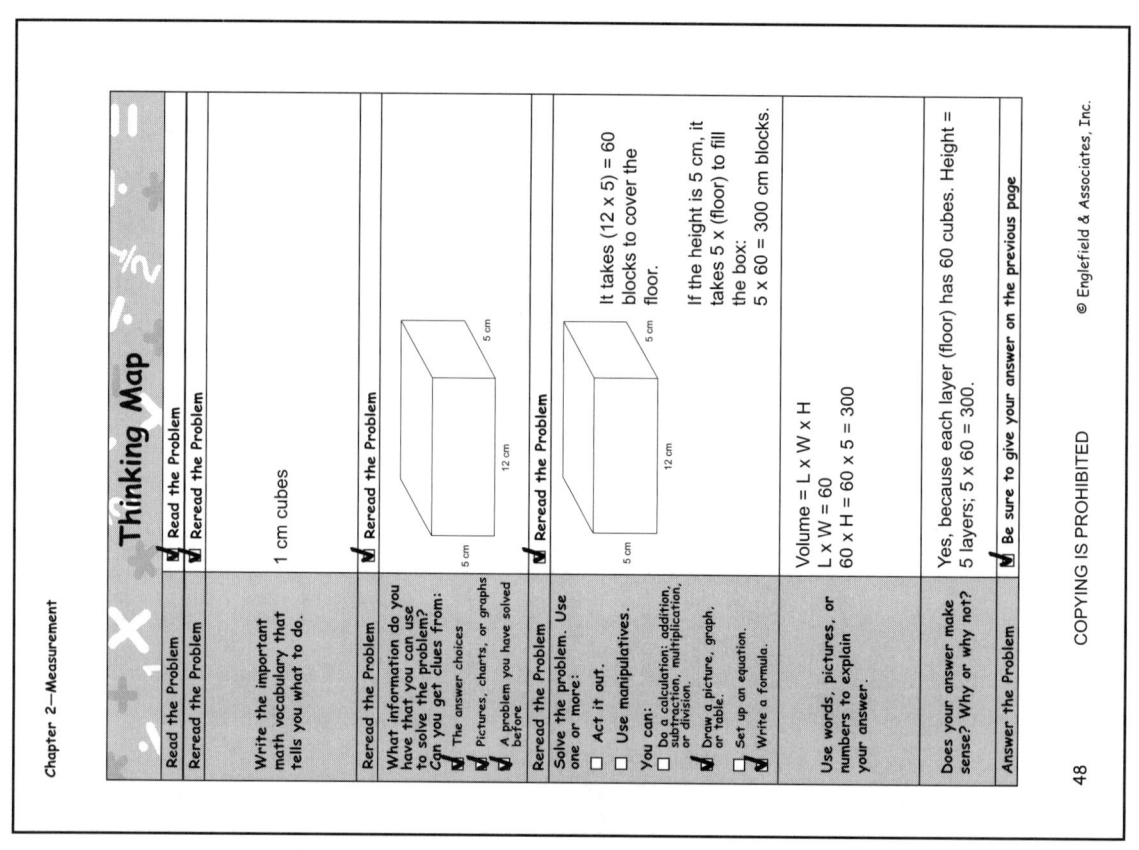

Read the Problem	☑ Read the Problem
Reread the Problem	☑ Reread the Problem
Write the important math vocabulary that tells you what to do.	1 cm cubes
Reread the Problem	☑ Reread the Problem
What information do you have that you can use to solve the problem? Can you get clues from: ☑ The answer choices ☑ Pictures, charts, or graphs ☑ A problem you have solved before	(box: 5 cm × 12 cm × 5 cm)
Reread the Problem	☑ Reread the Problem
Solve the problem. Use one or more: ☐ Act it out. ☐ Use manipulatives. You can: ☐ Do a calculation: addition, subtraction, multiplication, or division. ☑ Draw a picture, graph, or table. ☐ Set up an equation. ☑ Write a formula.	(box: 5 cm × 12 cm × 5 cm) It takes (12 × 5) = 60 blocks to cover the floor. If the height is 5 cm, it takes 5 × (floor) to fill the box: 5 × 60 = 300 cm blocks.
Use words, pictures, or numbers to explain your answer.	Volume = L × W × H L × W = 60 60 × H = 60 × 5 = 300
Does your answer make sense? Why or why not?	Yes, because each layer (floor) has 60 cubes. Height = 5 layers; 5 × 60 = 300.
Answer the Problem	☑ Be sure to give your answer on the previous page

48

In the Student Workbook, the Thinking Map is blank.
In the Parent/Teacher Edition, suggested ways for completing the Thinking Maps are presented.

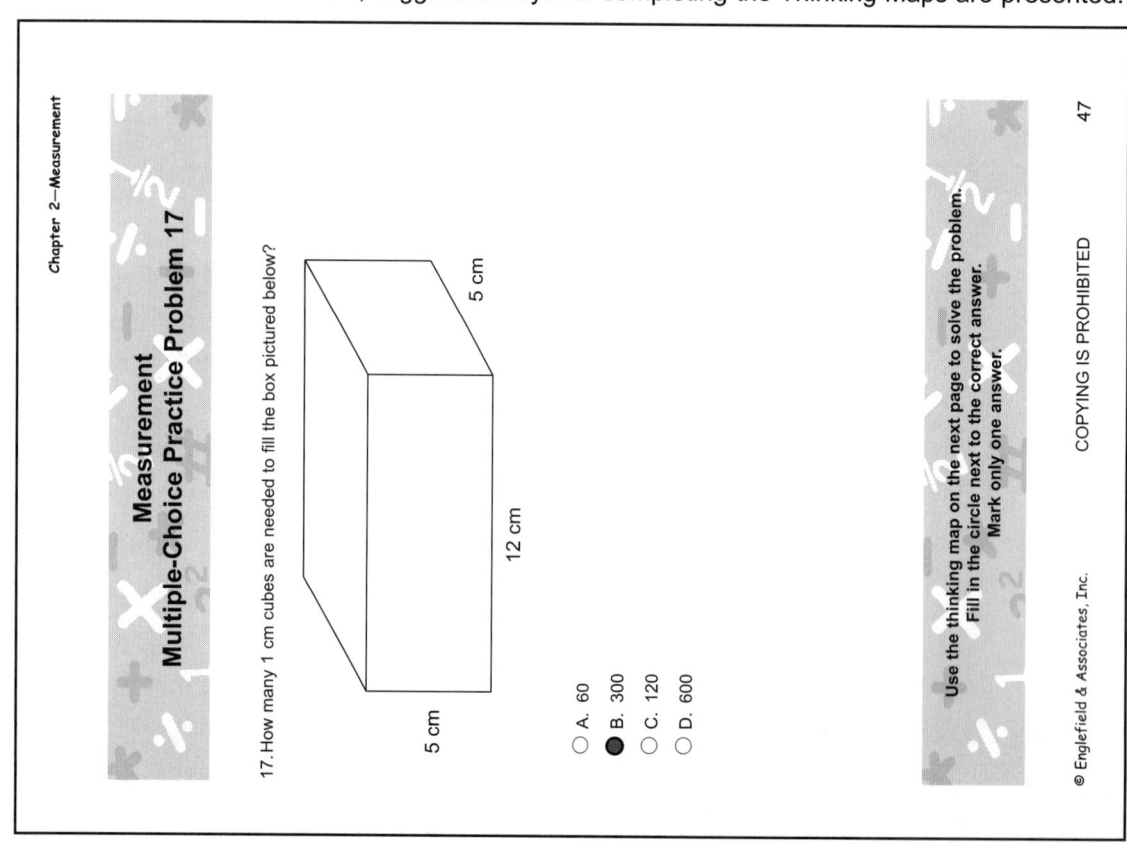

Measurement
Multiple-Choice Practice Problem 17

17. How many 1 cm cubes are needed to fill the box pictured below?

(box: 5 cm × 12 cm × 5 cm)

○ A. 60
● B. 300
○ C. 120
○ D. 600

Use the thinking map on the next page to solve the problem.
Fill in the circle next to the correct answer.
Mark only one answer.

47

50   COPYING IS PROHIBITED   ©Englefield & Associates, Inc.

Chapter 2—Measurement

Model Problem 5: Measurement Short-Answer

1. Prior Knowledge

Determine the student's prior knowledge about the strand, content of the problem, and/or mathematical process of the problem.

These questions can be used to help determine students' knowledge:

- What shapes of labels do you see on products?
- When do you have to match a net to a shape?

Additional prompts could be provided by using the following topics:

- designing labels to fit the shape of certain items
- experimenting with nets of various shapes
- advertising techniques
- creating art projects
- wrapping presents
- covering books

Chapter 2—Measurement

Model Problem 5: Measurement Short-Answer

5. La Monte was hired to design a new label for the ABC Soup Company. Describe the shape of the label and explain how it will fit the soup can.

1 point: The student describes the shape of the label as a rectangle.

1 point: The student describes the fit of the label as wrapping around the can or recognizes the label as a portion of the net of the can.

Use the thinking map on the next page to solve the problem. Write your answer in the box.

© Englefield & Associates, Inc.   COPYING IS PROHIBITED   49

Chapter 2—Measurement

2. Model

Model the use of the mathematical thinking map through class discussion with the thinking process modeled aloud as the thinking map is completed.

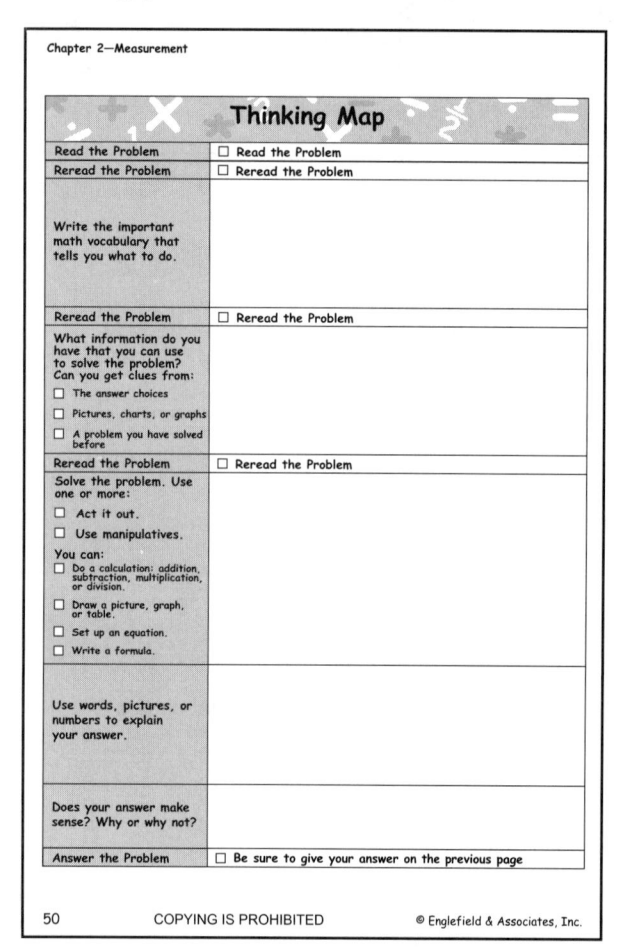

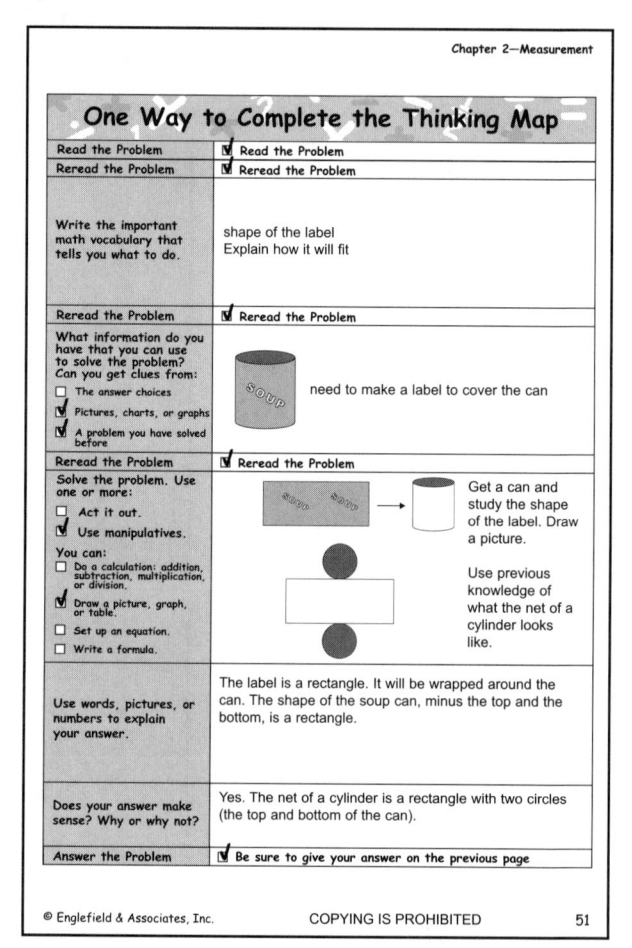

3. Guided Practice

Provide guided practice by presenting a problem for students to complete that matches the strand, content, or process you want to teach.

4. Independent Practice

Provide independent practice so students are provided with multiple-choice, short-answer, or extended-response problems.

Chapter 2—Measurement

Thinking Map

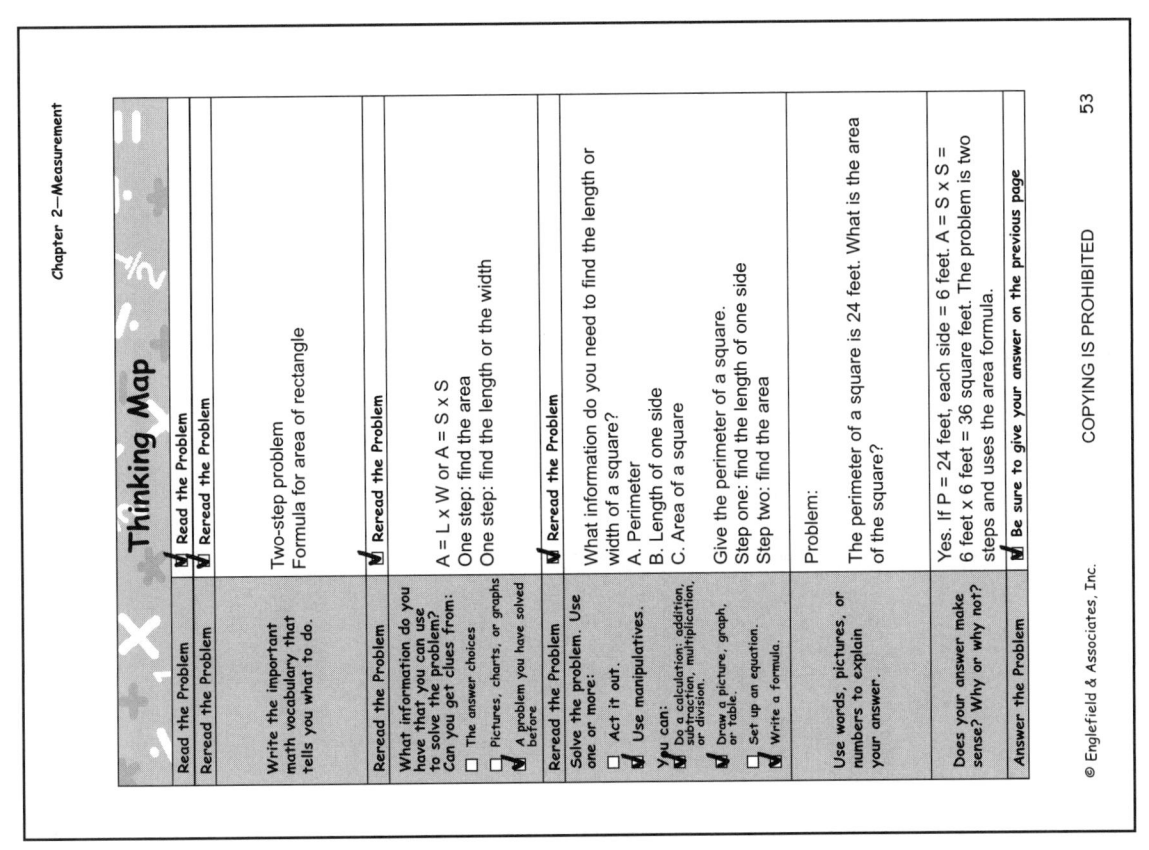

Read the Problem	☑ Read the Problem
	☑ Reread the Problem
Write the important math vocabulary that tells you what to do.	Two-step problem Formula for area of rectangle
Reread the Problem	☑ Reread the Problem
What information do you have that you can use to solve the problem? Can you get clues from: ☐ The answer choices ☐ Pictures, charts, or graphs ☑ A problem you have solved before	$A = L \times W$ or $A = S \times S$ One step: find the area One step: find the length or the width
Reread the Problem	☑ Reread the Problem
Solve the problem. Use one or more: ☐ Act it out ☐ Use manipulatives. You can: ☐ Do a calculation: addition, subtraction, multiplication, or division. ☐ Draw a picture, graph, or table. ☐ Set up an equation. ☑ Write a formula.	What information do you need to find the length or width of a square? A. Perimeter B. Length of one side C. Area of a square Give the perimeter of a square. Step one: find the length of one side Step two: find the area Problem: The perimeter of a square is 24 feet. What is the area of the square?
Use words, pictures, or numbers to explain your answer.	
Does your answer make sense? Why or why not?	Yes. If P = 24 feet, each side = 6 feet. $A = S \times S =$ 6 feet $\times$ 6 feet = 36 square feet. The problem is two steps and uses the area formula.
Answer the Problem	☑ Be sure to give your answer on the previous page

In the Student Workbook, the Thinking Map is blank.
In the Parent/Teacher Edition, suggested ways for completing the Thinking Maps are presented.

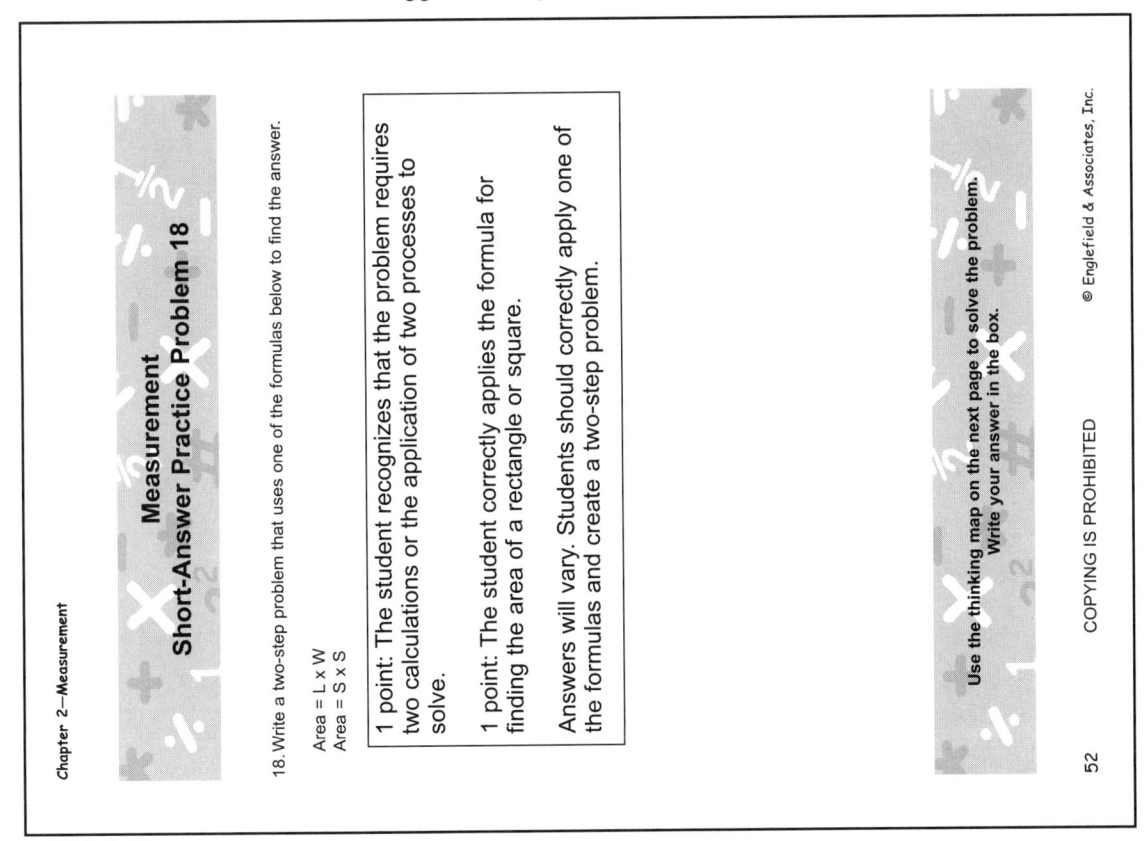

Measurement
Short-Answer Practice Problem 18

18. Write a two-step problem that uses one of the formulas below to find the answer.

Area = L x W
Area = S x S

1 point: The student recognizes that the problem requires two calculations or the application of two processes to solve.

1 point: The student correctly applies the formula for finding the area of a rectangle or square.

Answers will vary. Students should correctly apply one of the formulas and create a two-step problem.

Use the thinking map on the next page to solve the problem.
Write your answer in the box.

Chapter 2—Measurement

Measurement
Short-Answer Practice Problem 19

19. Eileen missed the problem shown below on her math test. What should she have written in the box to have the correct answer? Explain your answer.

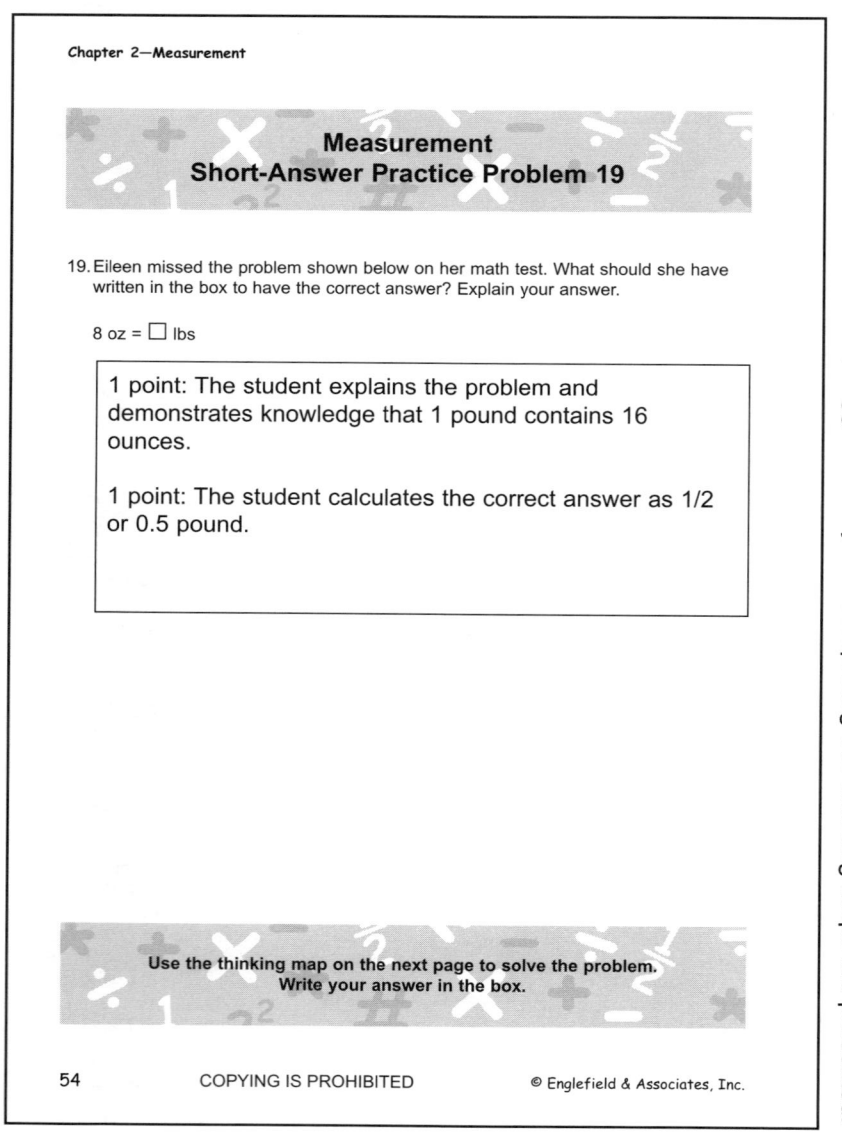

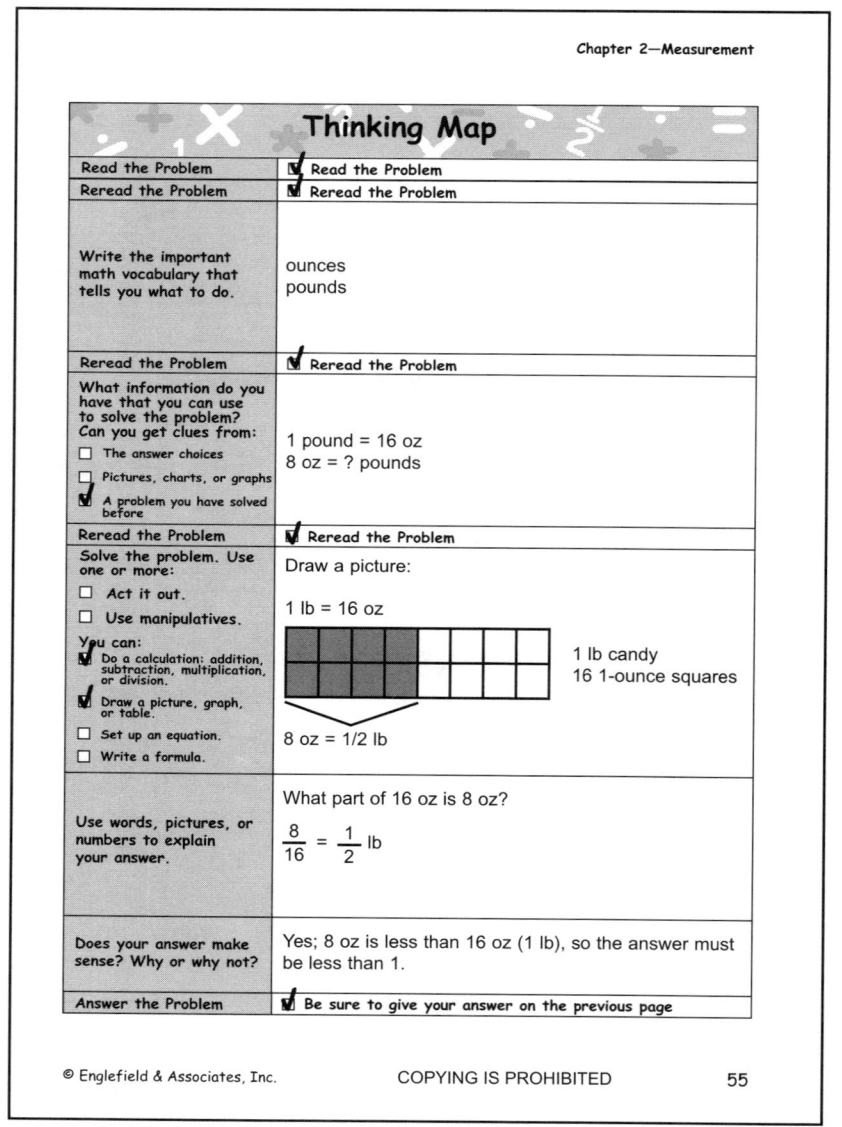

8 oz = ☐ lbs

> 1 point: The student explains the problem and demonstrates knowledge that 1 pound contains 16 ounces.
>
> 1 point: The student calculates the correct answer as 1/2 or 0.5 pound.

Use the thinking map on the next page to solve the problem. Write your answer in the box.

In the Student Workbook, the Thinking Map is blank.
In the Parent/Teacher Edition, suggested ways for completing the Thinking Maps are presented.

Thinking Map

Read the Problem	☑ Read the Problem
Reread the Problem	☑ Reread the Problem
Write the important math vocabulary that tells you what to do.	ounces pounds
Reread the Problem	☑ Reread the Problem
What information do you have that you can use to solve the problem? Can you get clues from: ☐ The answer choices ☐ Pictures, charts, or graphs ☑ A problem you have solved before	1 pound = 16 oz 8 oz = ? pounds
Reread the Problem	☑ Reread the Problem
Solve the problem. Use one or more: ☐ Act it out. ☐ Use manipulatives. You can: ☑ Do a calculation: addition, subtraction, multiplication, or division. ☑ Draw a picture, graph, or table. ☐ Set up an equation. ☐ Write a formula.	Draw a picture: 1 lb = 16 oz 1 lb candy 16 1-ounce squares 8 oz = 1/2 lb What part of 16 oz is 8 oz? $\frac{8}{16} = \frac{1}{2}$ lb
Use words, pictures, or numbers to explain your answer.	
Does your answer make sense? Why or why not?	Yes; 8 oz is less than 16 oz (1 lb), so the answer must be less than 1.
Answer the Problem	☑ Be sure to give your answer on the previous page

Model Problem 6: Measurement Extended-Response

1. Prior Knowledge

Determine the student's prior knowledge about the strand, content of the problem, and/or mathematical process of the problem.

These questions can be used to help determine students' knowledge:

- What do you know about time zones?
- Have you ever traveled to different time zones?
- What changes do you need to make when you change time zones?

Additional prompts could be provided by using the following topics:

- calling someone in a different time zone
- traveling
- ordering products from different countries
- news events

Model Problem 6: Measurement Extended-Response

6. The clocks below show the time zones in the United States. Marilyn lives in California, which is in the Pacific Standard Time Zone (PST). Marilyn's sister lives in the Eastern Standard Time Zone (EST) and leaves for work at 7:30 a.m. Marilyn would like to call her sister to wish her a happy birthday 15 minutes before her sister leaves for work. What time (in PST) must Marilyn call her sister?

PST   MST   CST   EST

1 point: The student determines that there is a three-hour time difference between PST and EST.

1 point: The student recognizes that the time in PST is earlier than the time in EST.

1 point: The student correctly identifies that Marilyn should call at 7:15 a.m. PST.

1 point: The student correctly identifies that 7:15 a.m. PST equates to 4:15 a.m. EST.

Use the thinking map on the next page to solve the problem. Write your answer in the box.

Chapter 2—Measurement

2. Model

Model the use of the mathematical thinking map through class discussion with the thinking process modeled aloud as the thinking map is completed.

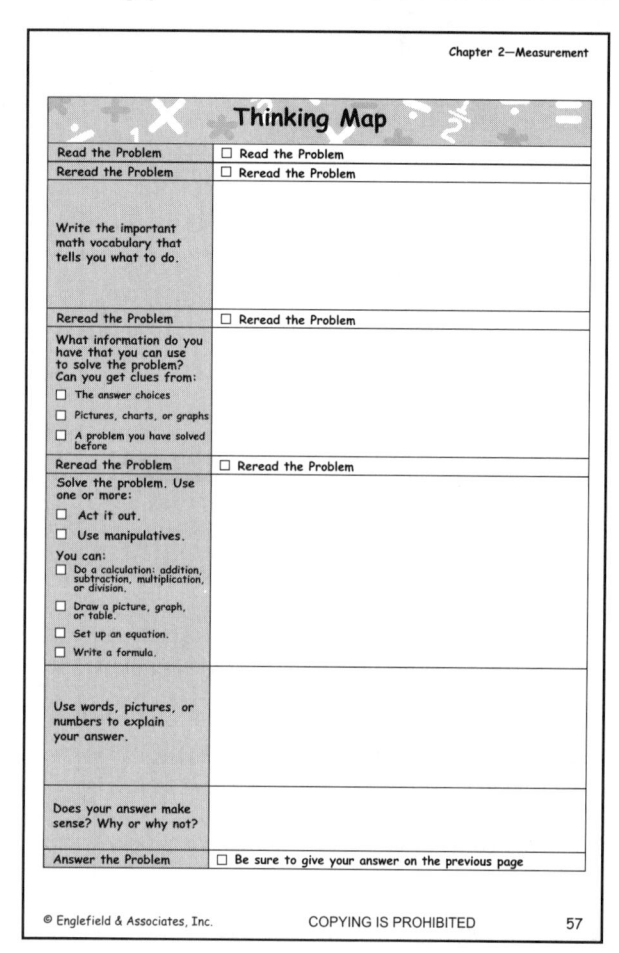

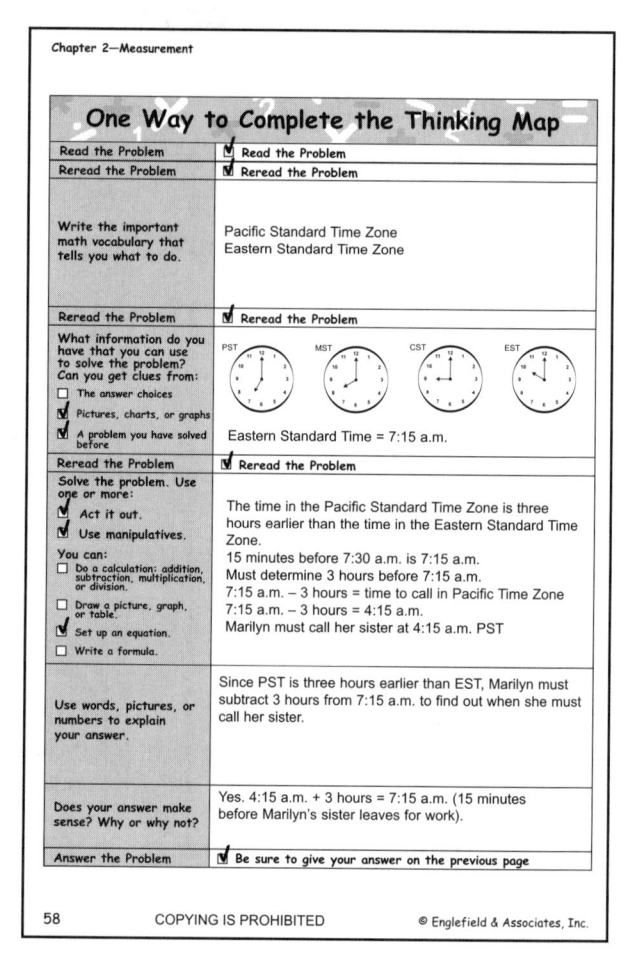

3. Guided Practice

Provide guided practice by presenting a problem for students to complete that matches the strand, content, or process you want to teach.

4. Independent Practice

Provide independent practice so students are provided with multiple-choice, short-answer, or extended-response problems.

Thinking Map

☑ Read the Problem	
☑ Reread the Problem	
Read the Problem	
Reread the Problem	
Write the important math vocabulary that tells you what to do.	Yards Cover the table 36 in wide
Reread the Problem	
What information do you have that you can use to solve the problem? Can you get clues from: ☐ The answer choices ☑ Pictures, charts, or graphs ☑ A problem you have solved before	Area = 18 sq ft One side = 3 ft
Reread the Problem	
Solve the problem. Use one or more: ☐ Act it out. ☐ Use manipulatives. You can: ☑ Do a calculation: addition, subtraction, multiplication, or division. ☑ Draw a picture, graph, or table. ☑ Set up an equation. ☐ Write a formula.	3 ft [18 sq ft] 3 ft A = L x W 3 x ? = 18; 18 ÷ 3 = 6 Side = 6 ft 6 ft = 2 yards 36 in = 3 ft
Use words, pictures, or numbers to explain your answer.	[18 sq ft] 36 in or 3 ft 6 ft or 2 yards Need 2 yards of fabric
Does your answer make sense? Why or why not?	Yes. One side = 6 ft; 6 ft = 2 yards.
Answer the Problem	
☑ Be sure to give your answer on the previous page	

60

In the Student Workbook, the Thinking Map is blank.
In the Parent/Teacher Edition, suggested ways for completing the Thinking Maps are presented.

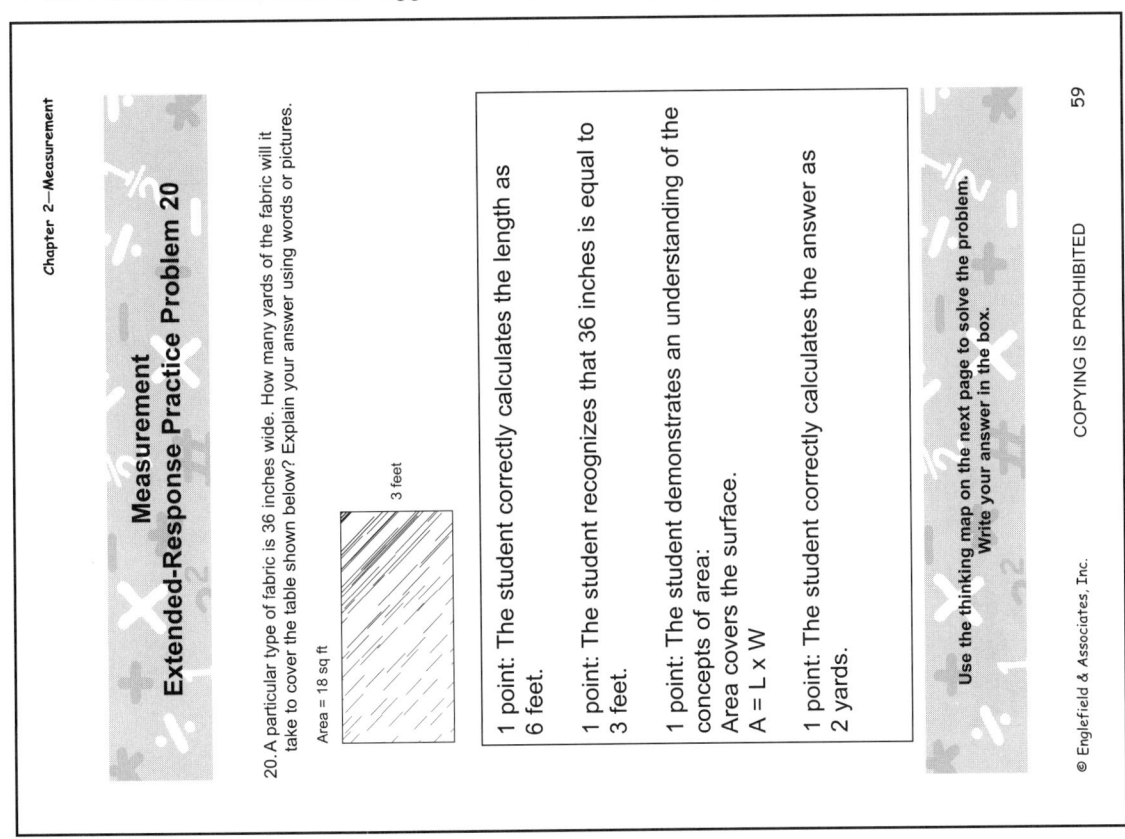

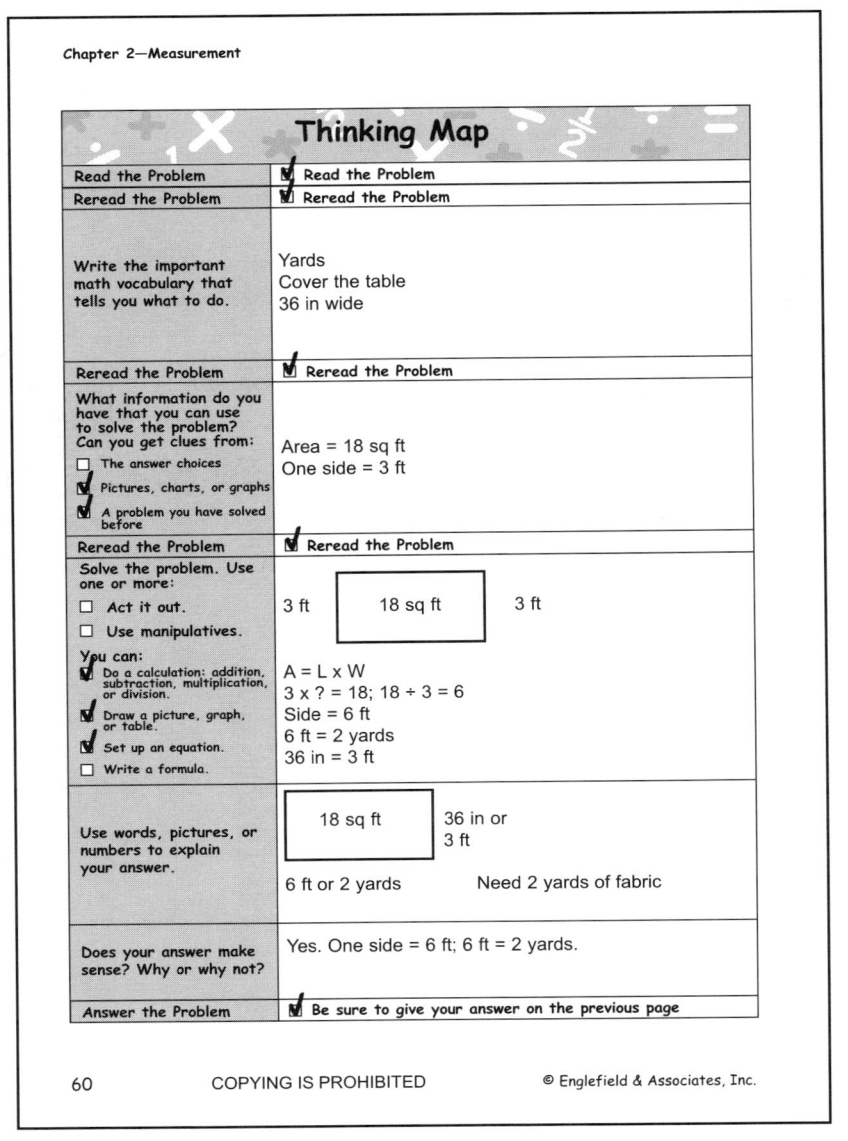

Measurement
Extended-Response Practice Problem 20

20. A particular type of fabric is 36 inches wide. How many yards of the fabric will it take to cover the table shown below? Explain your answer using words or pictures.

Area = 18 sq ft

3 feet

1 point: The student correctly calculates the length as 6 feet.

1 point: The student recognizes that 36 inches is equal to 3 feet.

1 point: The student demonstrates an understanding of the concepts of area:
Area covers the surface.
A = L x W

1 point: The student correctly calculates the answer as 2 yards.

Use the thinking map on the next page to solve the problem. Write your answer in the box.

Literature Connections for Measurement

Inch by Inch
Leo Lionni
© 1995

Measuring Penny
Loreen Leedy
© 2000

Millions to Measure
David M. Schwartz
© 2003

Hershey's Milk Chocolate Weights and Measures
Jerry Pallotta
© 2003

The Librarian Who Measured the Earth
Kathryn Lasky
© 1994

Me and the Measure of Things
Joan Sweeney
© 2001

Pigs on a Blanket
Amy Axelrod
© 1998

Telling Time with Big Mama Cat
Dan Harper
© 1998

How Big Is a Foot?
Rolf Myller and Susan McCrath
© 1991

Jim and the Beanstalk
Raymond Briggs
© 1997

Spaghetti and Meatballs for All: A Mathematical Story
Marilyn Burns and Debbie Tilley
© 1997

How Tall, How Short, How Faraway
David A. Adler
© 1999

Literature Connections for Measurement (Continued)

The Math Chef: Over 60 Math Activities and Recipes for Kids
Joan D'Amico and Karen Eich Drummond
www.carolina.com
© 1996

Weighing the Elephant
Ting-Xing Ye and Suzane Langlois
© 1999

Esio Trot
Roald Dahl
© 1990

The Borrowers
Mary Norton
© 1953

The Light Princess
George MacDonald
© 1977

There's No Place Like Space: All About Our Solar System
Tish Rabe
© 1999

What's Smaller than a Pygmy Shrew?
Robert E. Wells
© 1995

Wilma Unlimited: How Wilma Rudolph Became the World's Fastest Woman
Kathleen Krull
© 2000

How Big is a Foot?
Rolf Myller
© 1991

How Tall, How Short, How Far Away?
David Adler
© 2000

The Oxford Treasury of Time Poems
Michael Harrison and Christopher Stuart-Clark
© 1999

Fannie in the Kitchen: The Whole Story From Soup to Nuts of How Fannie Farmer Invented Recipes with Precise Measurements
Deborah Hopkinson
© 2001

Chapter 2—Measurement

For Future Reference

Note different strategies students used to solve the problems.

Chapter 3
Geometry

The purposes of this chapter include
1. Providing a definition of Geometry and what it means for fourth-grade students.
2. Offering teaching tips that reinforce Geometry concepts.
3. Offering multiple-choice, short-answer, and extended-response questions dealing with Geometry.
4. Providing ideas for connecting Geometry concepts and processes with other areas of the curriculum.

The following teaching tools are provided for extending students' thinking with Geometry:
- Model lesson for Geometry multiple-choice problems
- Seven Geometry multiple-choice practice problems
- Model lesson for Geometry short-answer problems
- Two Geometry short-answer practice problems
- Model lesson for Geometry extended-response problems
- One Geometry extended-response practice problem

What is Geometry?
Students are expected to model intersecting, parallel, and perpendicular lines. Students are expected to compare two- and three-dimensional objects. Students are expected to identify quadrilaterals and triangles based on angle measures and side lengths. Students are expected to describe points, lines, planes, and to plot ordered pairs on a coordinate plane. Students are expected to solve problems in other areas of mathematics using geometric models.

Chapter 3—Geometry

What does Geometry look like?

- Students will use manipulatives to model intersecting, parallel, and perpendicular lines.
- Students will compare two- and three-dimensional objects.
- Students will compare quadrilaterals.
- Students will identify triangles by their angle measurements and side lengths.
- Students will identify points, lines, and planes.
- Students will plot ordered pairs on a coordinate plane.
- Students will use reflections (flips), rotations (turns), and translations (slides) to solve problems.
- Students will use geometric models to solve a variety of problems.

Vocabulary related to Geometry:

Acute angle: An angle measuring less than 90°.

Angle: The amount of turning between two lines meeting at a common point. Some common types of angles are: right angle = 90°; acute angle = less than 90°; obtuse angle = between 90° and 180°; straight angle = 180°; reflex angle = between 180° and 360°; revolution = 360°.

Center: A point that is the same distance from all points on the circumference of a circle or on a circle of a sphere; a point that is the same distance from all of the vertices of a regular polygon; a middle position.

Circle: A plane shape bounded by a continuous line which is always the same distance from the center.

Circumference: The distance around a circle; the boundary or perimeter of a circle. The formula for circumference is $C = 2\pi r = 2 \times 3.14 \times radius$.

Congruent: Having the same size and shape.

Coordinate grid system: A two-dimensional system in which the coordinates of a point are its distance from the two intersecting, usually perpendicular, lines called axes.

Diameter: A straight line passing through the center of a circle to touch both sides of the circumference.

Dimension: Refers to length, width, or height.

Equilateral triangle: A triangle with three equal angles and three equal sides.

Face: A flat surface of a three-dimensional shape.

Graph: A drawing or diagram used to record information.

Intersecting lines: Two lines that cross over each other at exactly one point.

Line: A path of points that extends continuously in two opposing directions.

Line segment: Part of a line.

Map: To establish the coordinates of an element in a set; to plan, in detail, the path to get from one point to another.

Negative number: A number less than zero; use "-" to show a negative number, e.g., -20, -4.

Obtuse angle: An angle measuring between 90° and 180°.

Parallel lines: Lines in the same plane that do not cross; the distance between the lines is constant.

Perpendicular: At right angles to the horizon; lines that intersect at right angles to each other.

Perpendicular lines: Lines that intersect each other at a right (90°) angle.

Plane: A flat surface.

Polygon: A shape that has three or more straight sides.

Point: A position in space.

Quadrilateral: A polygon with four sides such as a square, a rectangle, a parallelogram, or a trapezoid.

Radius: The distance from the center of a circle to its outer edge.

Ray: A line that has a starting point but no end point (goes on indefinitely).

Reflection (flip): A mirror view.

Right angle: An angle measuring 90°.

Right-angled triangle: A triangle with one right (90°) angle; also called a right triangle.

Rotation (turn): A transformation in which an object is turned; a transformation about a fixed point such that every point in the object turns through the same angle relative to that fixed point.

Scalene triangle: A triangle with no equal angles or equal sides.

Side: A line bounding a plane figure, a surface bounding a solid figure, or a line or curve on the edge of a shape.

Symmetry: Correspondence of form and configuration on opposite sides of a dividing line or plane or about a center or an axis.

Three-dimensional: Describes a shape that has three dimensions—length, width, and height.

Translation (slide): A transformation in which an item is moved in any direction without being rotated.

Triangle: A polygon with three angles and three sides.

Two-dimensional: Describes a shape that has two dimensions, usually length and width or length and height.

Vertex (Vertices—plural): The point where lines meet; corner.

Chapter 3—Geometry

Teaching Tips

- Provide the specific vocabulary for describing and measuring.
- Remember that visual/spatial problems may pose difficulties for students as they work problems with three-dimensional objects or those that require visualization.
- Model visualization techniques. Some students may need teachers to take them from concrete to abstract using tangible objects or representations.
- Students with fine motor problems may have difficulty sketching figures.
- Demonstrate how geometry can be used in authentic, real-life situations.

For Future Reference

Note different strategies students used to solve the problems.

Model Problem 7: Geometry Multiple-Choice

1. Prior Knowledge

Determine the student's prior knowledge about the strand, content of the problem, and/or mathematical process of the problem.

These questions can be used to help determine students' knowledge:

- What are all the ways that you can estimate a measurement?
- What do you know about units of measure?
- How do you use real objects to estimate measurements?

Additional prompts could be provided by using the following topics:

- designs in art
- sewing projects
- building projects (LEGO® bricks, a birdhouse, etc.)

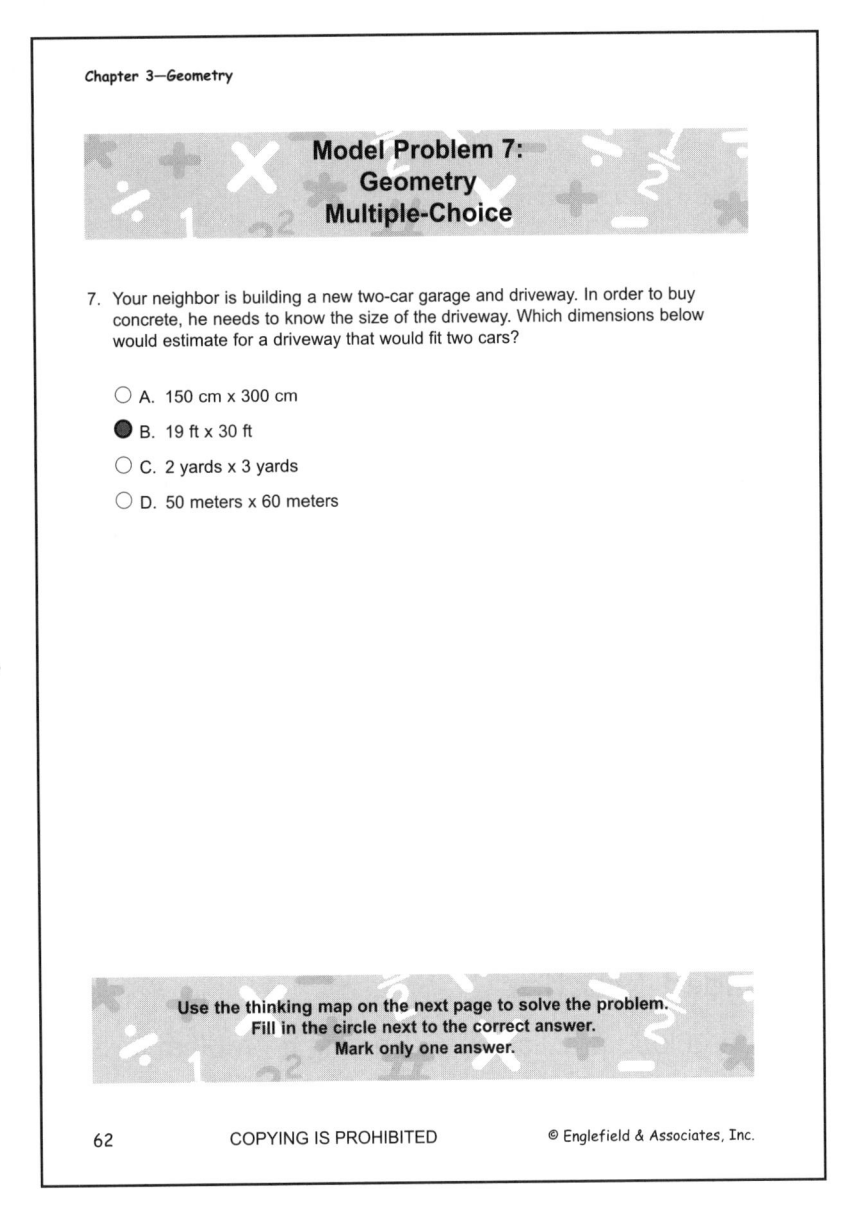

Model Problem 7: Geometry Multiple-Choice

7. Your neighbor is building a new two-car garage and driveway. In order to buy concrete, he needs to know the size of the driveway. Which dimensions below would estimate for a driveway that would fit two cars?

    ○ A. 150 cm x 300 cm
    ● B. 19 ft x 30 ft
    ○ C. 2 yards x 3 yards
    ○ D. 50 meters x 60 meters

Use the thinking map on the next page to solve the problem.
Fill in the circle next to the correct answer.
Mark only one answer.

Chapter 3—Geometry

2. Model

Model the use of the mathematical thinking map through class discussion with the thinking process modeled aloud as the thinking map is completed.

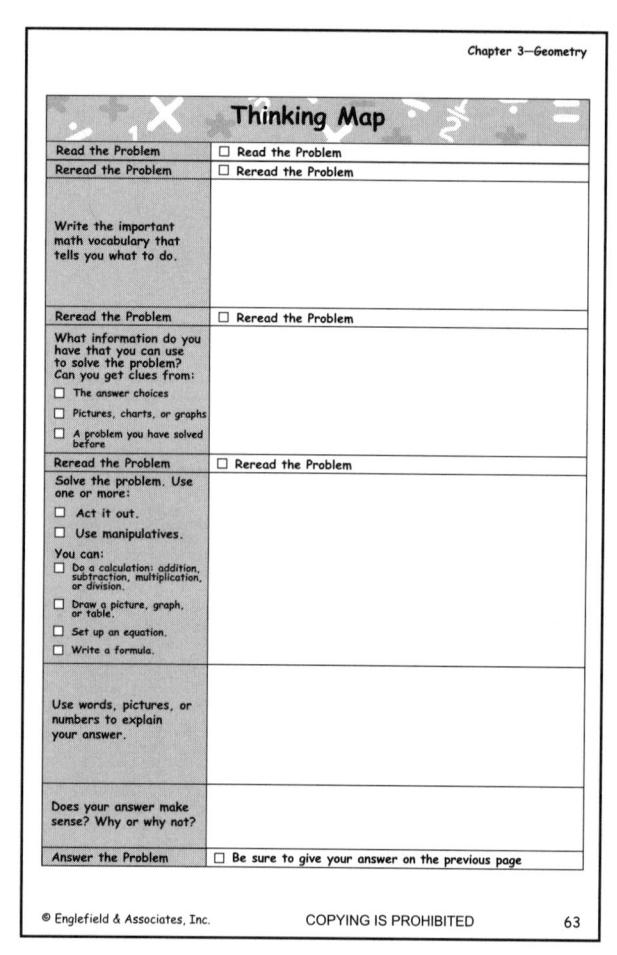

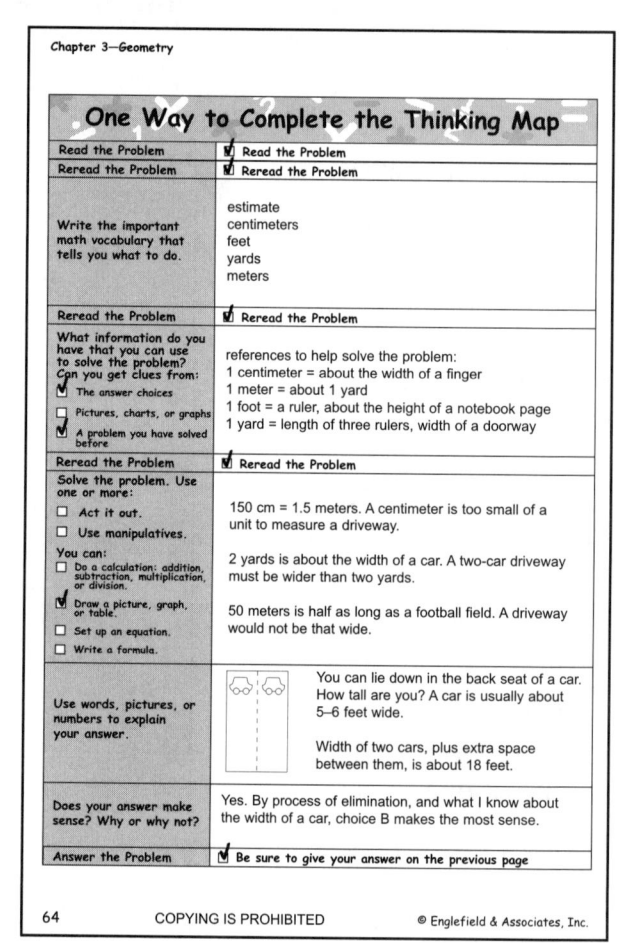

Thinking Map

Read the Problem	☐ Read the Problem
Reread the Problem	☐ Reread the Problem
Write the important math vocabulary that tells you what to do.	
Reread the Problem	☐ Reread the Problem
What information do you have that you can use to solve the problem? Can you get clues from: ☐ The answer choices ☐ Pictures, charts, or graphs ☐ A problem you have solved before	
Reread the Problem	☐ Reread the Problem
Solve the problem. Use one or more: ☐ Act it out. ☐ Use manipulatives. You can: ☐ Do a calculation: addition, subtraction, multiplication, or division. ☐ Draw a picture, graph, or table. ☐ Set up an equation. ☐ Write a formula.	
Use words, pictures, or numbers to explain your answer.	
Does your answer make sense? Why or why not?	
Answer the Problem	☐ Be sure to give your answer on the previous page

One Way to Complete the Thinking Map

Read the Problem	☑ Read the Problem
Reread the Problem	☑ Reread the Problem
Write the important math vocabulary that tells you what to do.	estimate centimeters feet yards meters
Reread the Problem	☑ Reread the Problem
What information do you have that you can use to solve the problem? Can you get clues from: ☑ The answer choices ☐ Pictures, charts, or graphs ☑ A problem you have solved before	references to help solve the problem: 1 centimeter = about the width of a finger 1 meter = about 1 yard 1 foot = a ruler, about the height of a notebook page 1 yard = length of three rulers, width of a doorway
Reread the Problem	☑ Reread the Problem
Solve the problem. Use one or more: ☐ Act it out. ☐ Use manipulatives. You can: ☐ Do a calculation: addition, subtraction, multiplication, or division. ☑ Draw a picture, graph, or table. ☐ Set up an equation. ☐ Write a formula.	150 cm = 1.5 meters. A centimeter is too small of a unit to measure a driveway. 2 yards is about the width of a car. A two-car driveway must be wider than two yards. 50 meters is half as long as a football field. A driveway would not be that wide.
Use words, pictures, or numbers to explain your answer.	You can lie down in the back seat of a car. How tall are you? A car is usually about 5–6 feet wide. Width of two cars, plus extra space between them, is about 18 feet.
Does your answer make sense? Why or why not?	Yes. By process of elimination, and what I know about the width of a car, choice B makes the most sense.
Answer the Problem	☑ Be sure to give your answer on the previous page

3. Guided Practice

Provide guided practice by presenting a problem for students to complete that matches the strand, content, or process you want to teach.

4. Independent Practice

Provide independent practice so students are provided with multiple-choice, short-answer, or extended-response problems.

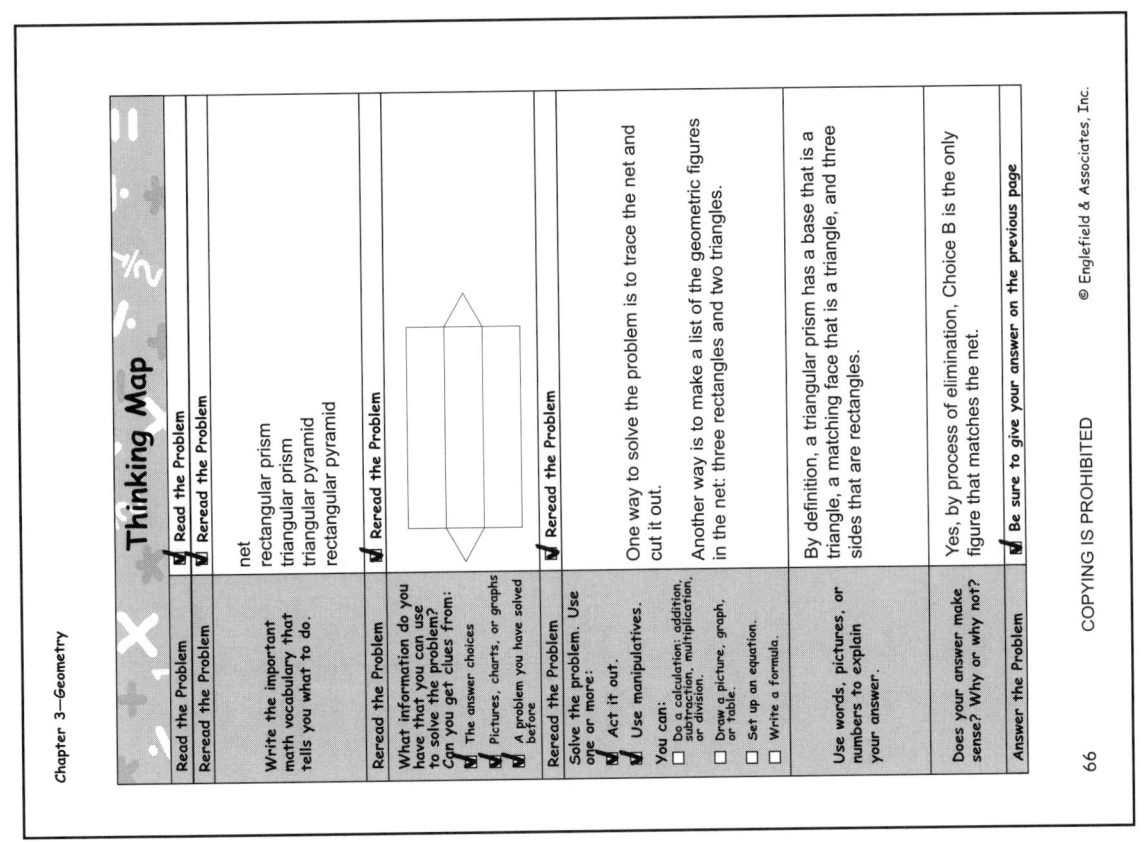

Chapter 3—Geometry

Thinking Map

☑ Read the Problem	
☑ Reread the Problem	
Write the important math vocabulary that tells you what to do.	net rectangular prism triangular prism triangular pyramid rectangular pyramid
☑ Reread the Problem	
What information do you have that you can use to solve the problem? Can you get clues from: ☑ The answer choices ☑ Pictures, charts, or graphs ☑ A problem you have solved before	
☑ Reread the Problem	
Solve the problem. Use one or more: ☑ Act it out. ☑ Use manipulatives. You can: ☐ Do a calculation: addition, subtraction, multiplication, or division. ☐ Draw a picture, graph, or table. ☐ Set up an equation. ☐ Write a formula.	One way to solve the problem is to trace the net and cut it out. Another way is to make a list of the geometric figures in the net: three rectangles and two triangles.
Use words, pictures, or numbers to explain your answer.	By definition, a triangular prism has a base that is a triangle, a matching face that is a triangle, and three sides that are rectangles.
Does your answer make sense? Why or why not?	Yes, by process of elimination, Choice B is the only figure that matches the net.
Answer the Problem	☑ Be sure to give your answer on the previous page

66 COPYING IS PROHIBITED © Englefield & Associates, Inc.

In the Student Workbook, the Thinking Map is blank.
In the Parent/Teacher Edition, suggested ways for completing the Thinking Maps are presented.

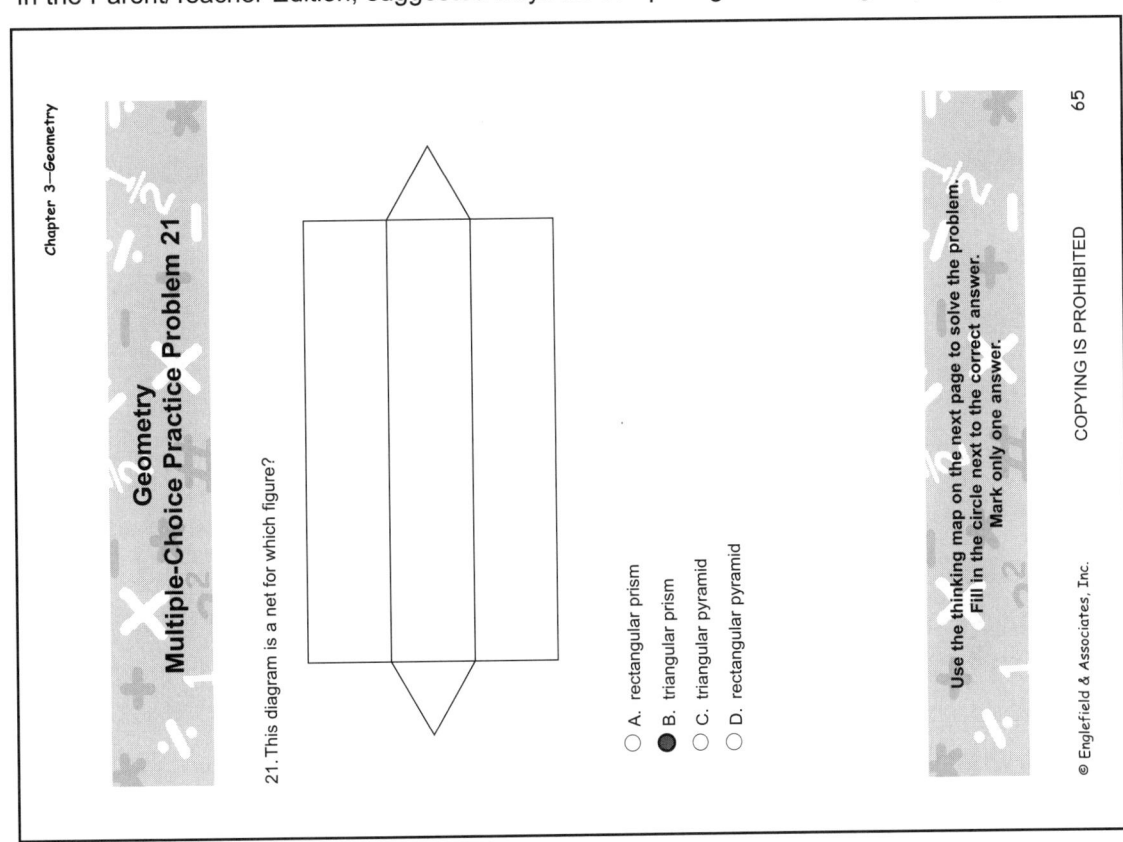

Geometry
Multiple-Choice Practice Problem 21

21. This diagram is a net for which figure?

○ A. rectangular prism
● B. triangular prism
○ C. triangular pyramid
○ D. rectangular pyramid

Use the thinking map on the next page to solve the problem.
Fill in the circle next to the correct answer.
Mark only one answer.

© Englefield & Associates, Inc. COPYING IS PROHIBITED

Thinking Map

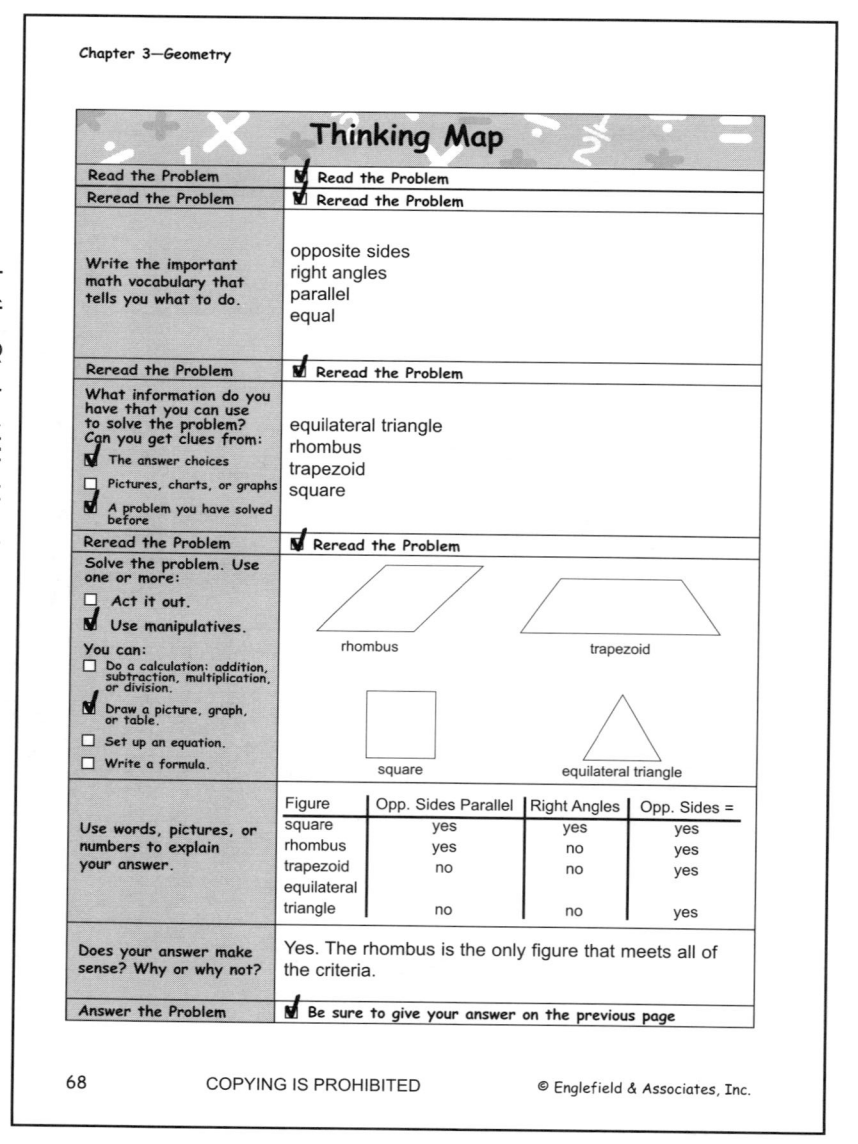

Read the Problem	☑ Read the Problem
	☑ Reread the Problem
Write the important math vocabulary that tells you what to do.	opposite sides right angles parallel equal
Reread the Problem	☑ Reread the Problem
What information do you have that you can use to solve the problem? Can you get clues from:	equilateral triangle rhombus trapezoid square
☐ The answer choices ☑ Pictures, charts, or graphs ☐ A problem you have solved before	
Reread the Problem	☑ Reread the Problem
Solve the problem. Use one or more: ☐ Act it out. ☑ Use manipulatives. You can: ☐ Do a calculation: addition, subtraction, multiplication, or division. ☑ Draw a picture, graph, or table. ☐ Set up an equation. ☐ Write a formula.	rhombus trapezoid square equilateral triangle
Use words, pictures, or numbers to explain your answer.	Figure | Opp. Sides Parallel | Right Angles | Opp. Sides = square | yes | yes | yes rhombus | yes | no | yes trapezoid | no | no | no equilateral triangle | — | — | yes
Does your answer make sense? Why or why not?	Yes. The rhombus is the only figure that meets all of the criteria.
Answer the Problem	☑ Be sure to give your answer on the previous page

In the Student Workbook, the Thinking Map is blank.
In the Parent/Teacher Edition, suggested ways for completing the Thinking Maps are presented.

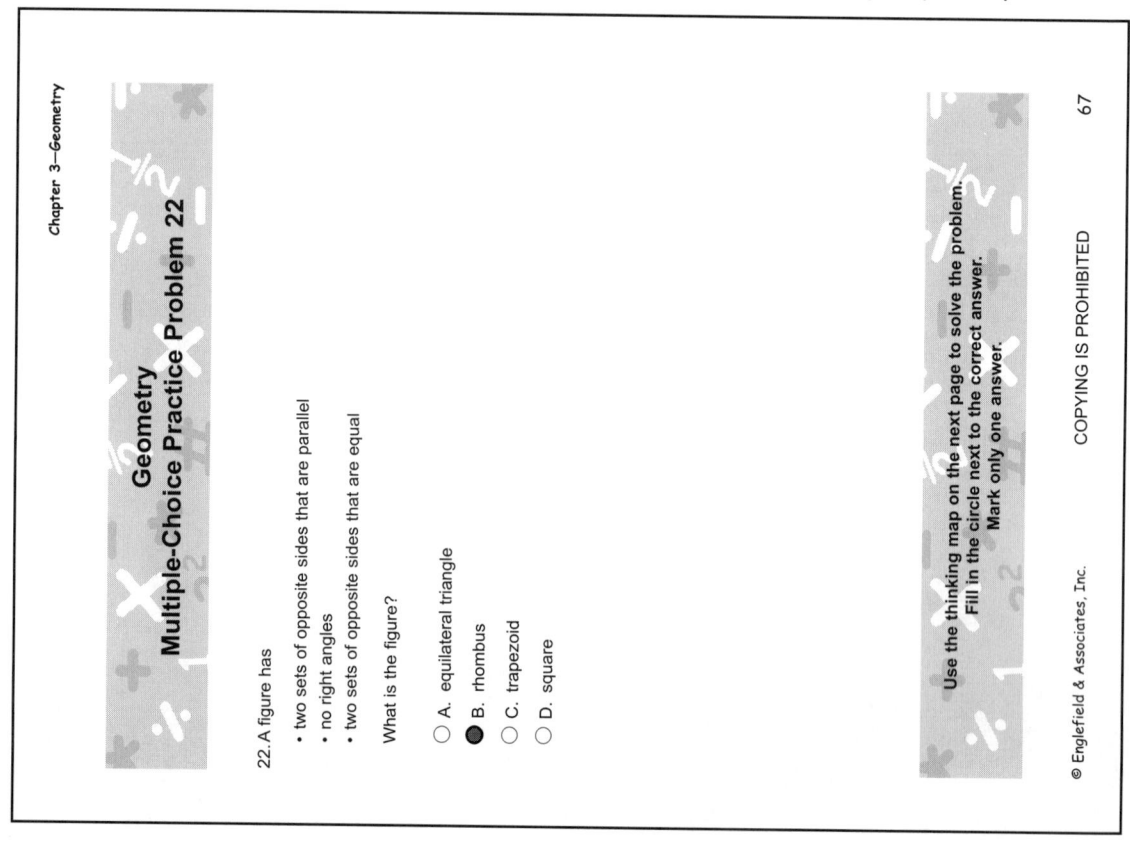

Geometry
Multiple-Choice Practice Problem 22

22. A figure has
- two sets of opposite sides that are parallel
- no right angles
- two sets of opposite sides that are equal

What is the figure?

○ A. equilateral triangle
● B. rhombus
○ C. trapezoid
○ D. square

Use the thinking map on the next page to solve the problem.
Fill in the circle next to the correct answer.
Mark only one answer.

Chapter 3—Geometry

Thinking Map

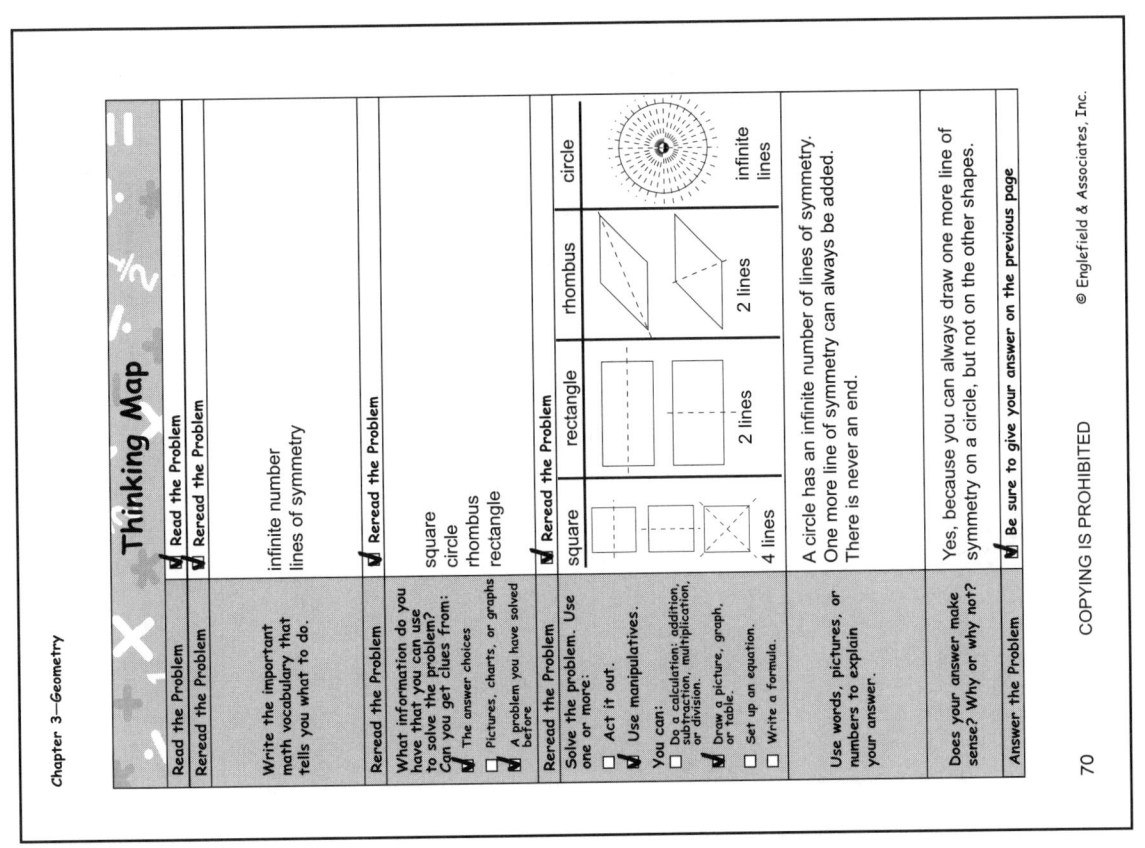

Read the Problem	☑ Read the Problem ☑ Reread the Problem
Write the important math vocabulary that tells you what to do.	infinite number lines of symmetry
Reread the Problem	☑ Reread the Problem
What information do you have that you can use to solve the problem? Can you get clues from: ☐ The answer choices ☐ Pictures, charts, or graphs ☐ A problem you have solved before	square circle rhombus rectangle
Reread the Problem	☑ Reread the Problem
Solve the problem. Use one or more: ☐ Act it out. ☐ Use manipulatives. You can: ☐ Do a calculation: addition, subtraction, multiplication, or division. ☐ Draw a picture, graph, or table. ☐ Set up an equation. ☐ Write a formula.	square — 4 lines rectangle — 2 lines rhombus — 2 lines circle — infinite lines A circle has an infinite number of lines of symmetry. One more line of symmetry can always be added. There is never an end.
Use words, pictures, or numbers to explain your answer.	
Does your answer make sense? Why or why not?	Yes, because you can always draw one more line of symmetry on a circle, but not on the other shapes.
Answer the Problem	☑ Be sure to give your answer on the previous page

In the Student Workbook, the Thinking Map is blank.
In the Parent/Teacher Edition, suggested ways for completing the Thinking Maps are presented.

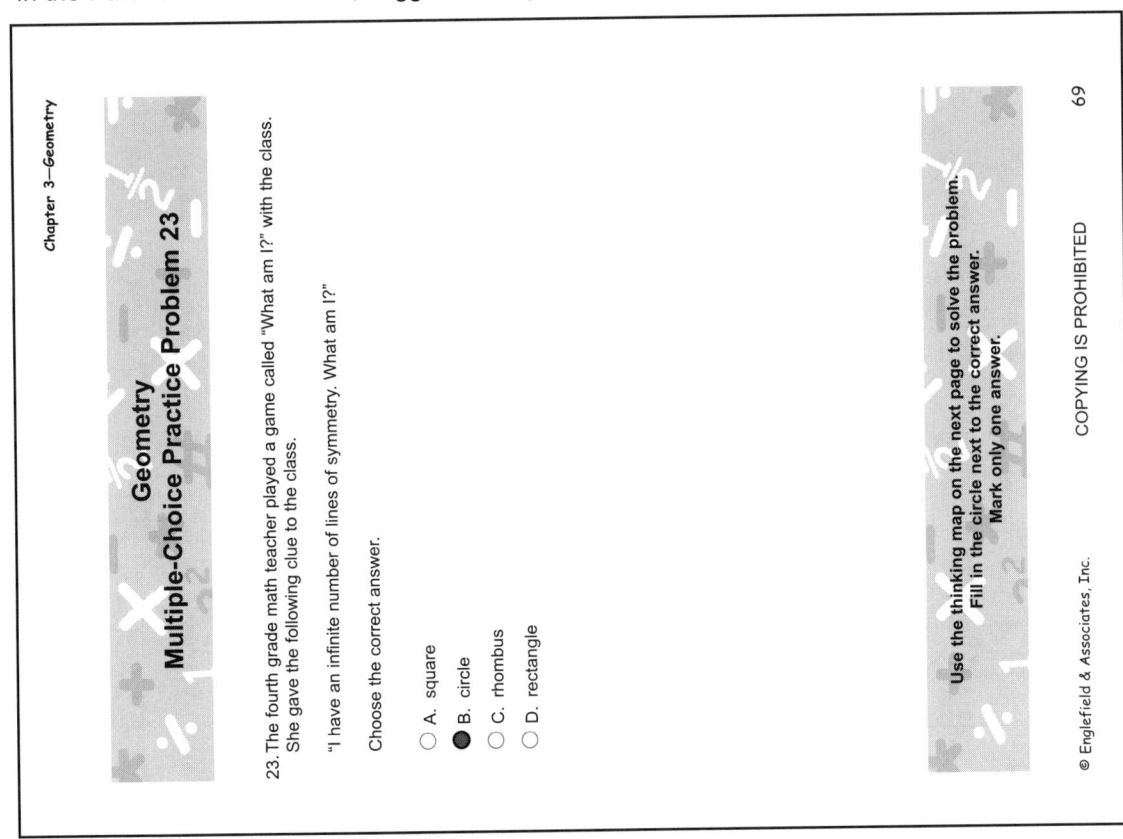

Geometry
Multiple-Choice Practice Problem 23

23. The fourth grade math teacher played a game called "What am I?" with the class. She gave the following clue to the class.

"I have an infinite number of lines of symmetry. What am I?"

Choose the correct answer.

○ A. square
● B. circle
○ C. rhombus
○ D. rectangle

Use the thinking map on the next page to solve the problem.
Fill in the circle next to the correct answer.
Mark only one answer.

Chapter 3—Geometry

Thinking Map

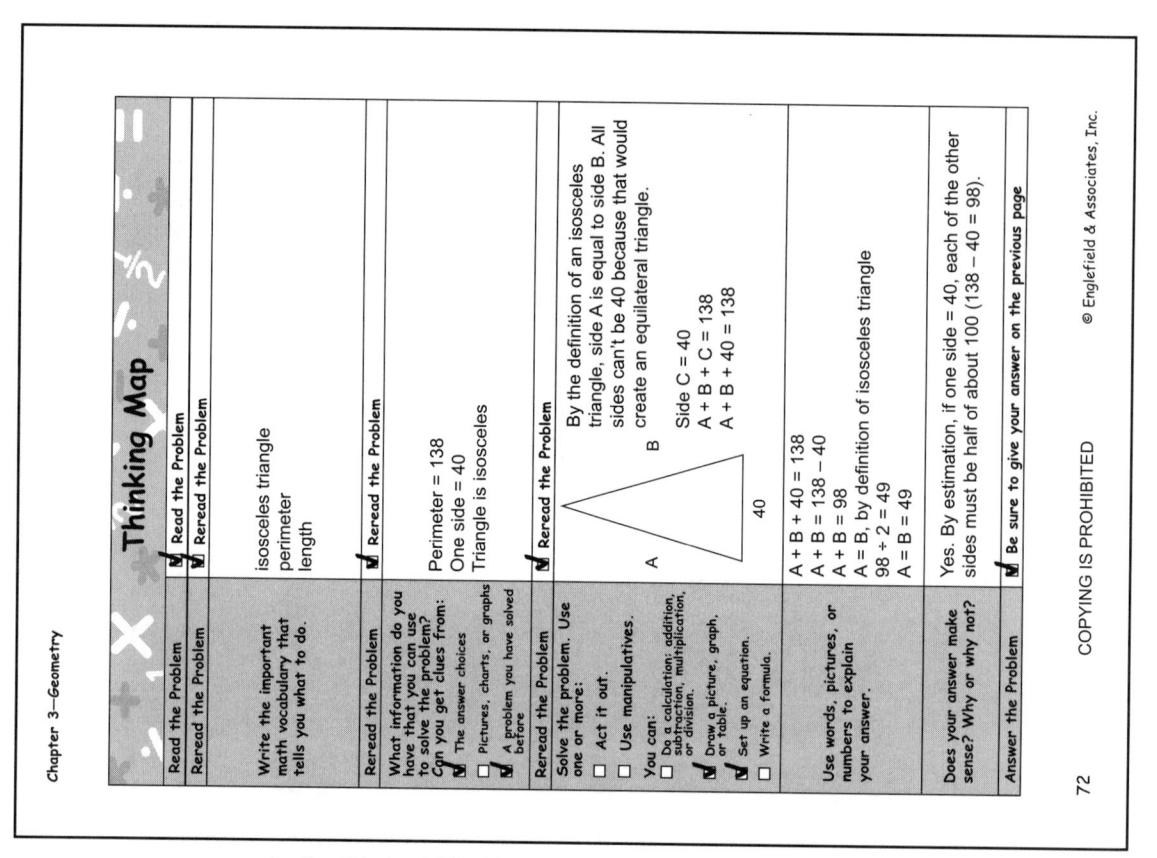

☑ Read the Problem	
☑ Reread the Problem	isosceles triangle perimeter length
Write the important math vocabulary that tells you what to do.	
☑ Reread the Problem	Perimeter = 138 One side = 40 Triangle is isosceles
What information do you have that you can use to solve the problem? Can you get clues from: ☑ The answer choices ☐ Pictures, charts, or graphs ☑ A problem you have solved before	
☑ Reread the Problem	By the definition of an isosceles triangle, side A is equal to side B. All sides can't be 40 because that would create an equilateral triangle. Side C = 40 A + B + C = 138 A + B + 40 = 138
Solve the problem. Use one or more: ☐ Act it out. ☐ Use manipulatives. You can: ☐ Do a calculation: addition, subtraction, multiplication, or division. ☑ Draw a picture, graph, or table. ☑ Set up an equation. ☐ Write a formula.	A + B + 40 = 138 A + B = 138 − 40 A + B = 98 A = B, by definition of isosceles triangle 98 ÷ 2 = 49 A = B = 49
Use words, pictures, or numbers to explain your answer.	
Does your answer make sense? Why or why not?	Yes. By estimation, if one side = 40, each of the other sides must be half of about 100 (138 − 40 = 98).
Answer the Problem	☑ Be sure to give your answer on the previous page

In the Student Workbook, the Thinking Map is blank.
In the Parent/Teacher Edition, suggested ways for completing the Thinking Maps are presented.

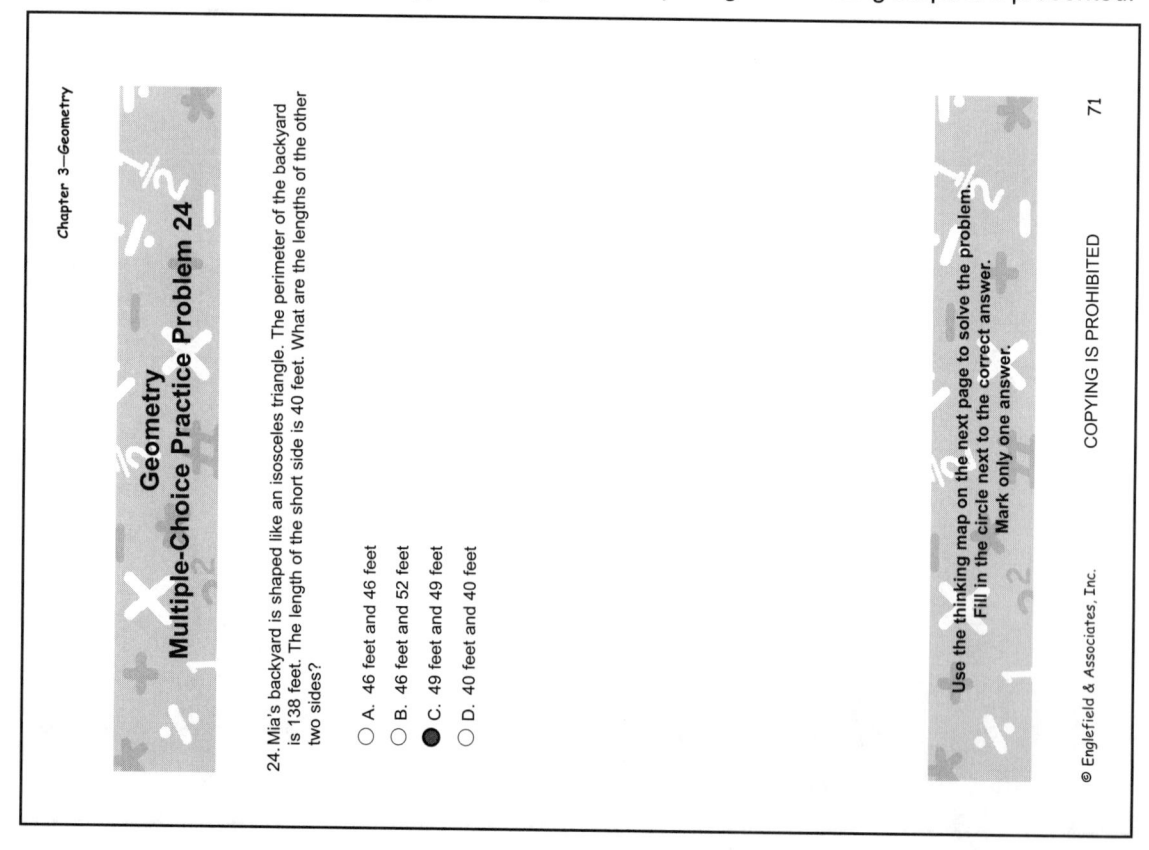

Chapter 3—Geometry

Geometry
Multiple-Choice Practice Problem 24

24. Mia's backyard is shaped like an isosceles triangle. The perimeter of the backyard is 138 feet. The length of the short side is 40 feet. What are the lengths of the other two sides?

○ A. 46 feet and 46 feet
○ B. 46 feet and 52 feet
● C. 49 feet and 49 feet
○ D. 40 feet and 40 feet

Use the thinking map on the next page to solve the problem.
Fill in the circle next to the correct answer.
Mark only one answer.

Thinking Map

Read the Problem
☑ Read the Problem
☑ Reread the Problem

Reread the Problem
Write the important math vocabulary that tells you what to do.

coordinate
congruent
triangle

Reread the Problem
☑ Reread the Problem

coordinates of m:
(3, 3), (6, 3), (3, 6)

Reread the Problem
What information do you have that you can use to solve the problem? Can you get clues from:
☑ The answer choices
☑ Pictures, charts, or graphs
☑ A problem you have solved before

Reread the Problem
Solve the problem. Use one or more:
☐ Act it out.
☐ Use manipulatives.
You can:
☐ Do a calculation: addition, subtraction, multiplication, or division.
☑ Draw a picture, graph, or table.
☐ Set up an equation.
☐ Write a formula.

Subtract x- and y-coordinates:
(9, 5) − (3, 3) = 6, 2
(12, 5) − (6, 3) = 6, 2
(9, 8) − (3, 6) = 6, 2
Triangle D is congruent.

Does your answer make sense? Why or why not?
Yes, figure D looks like it is congruent when compared to the original. Figures A, B, and C do not.

Answer the Problem
☑ Be sure to give your answer on the previous page.

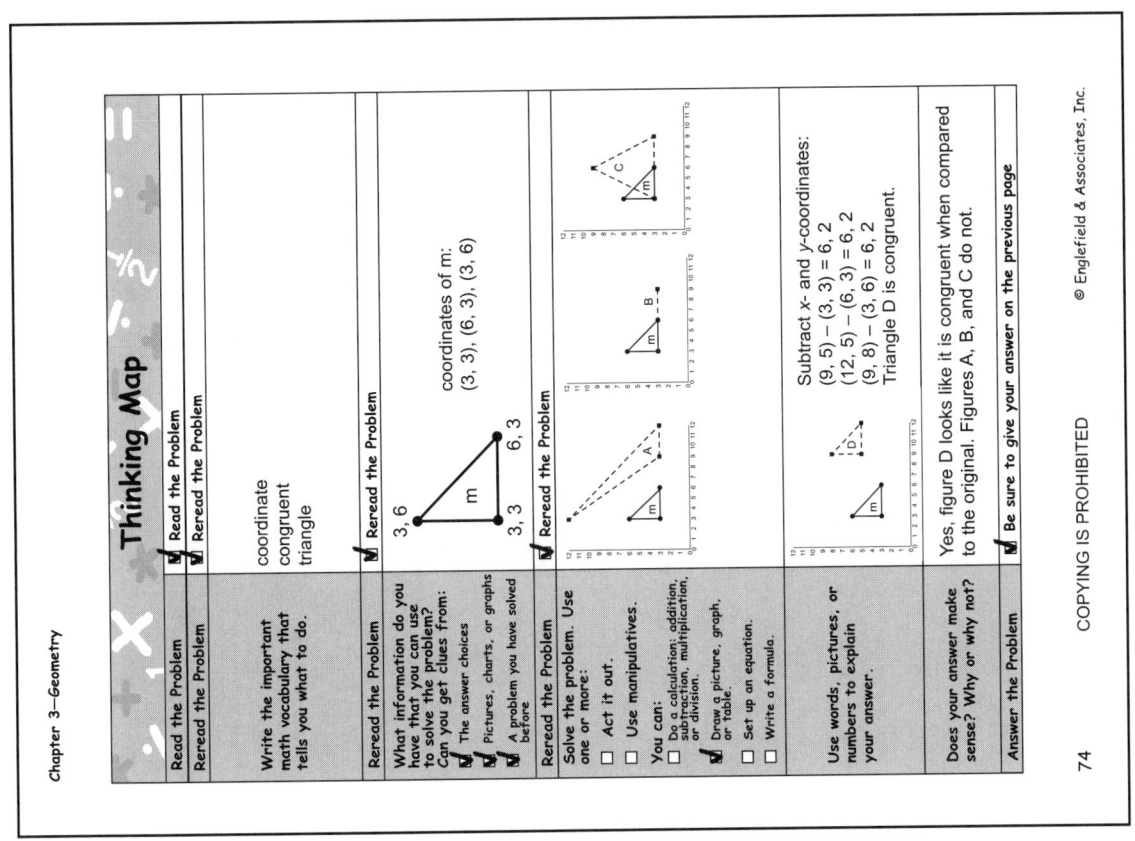

In the Student Workbook, the Thinking Map is blank.
In the Parent/Teacher Edition, suggested ways for completing the Thinking Maps are presented.

Geometry
Multiple-Choice Practice Problem 25

25. What are the coordinates of a congruent triangle?

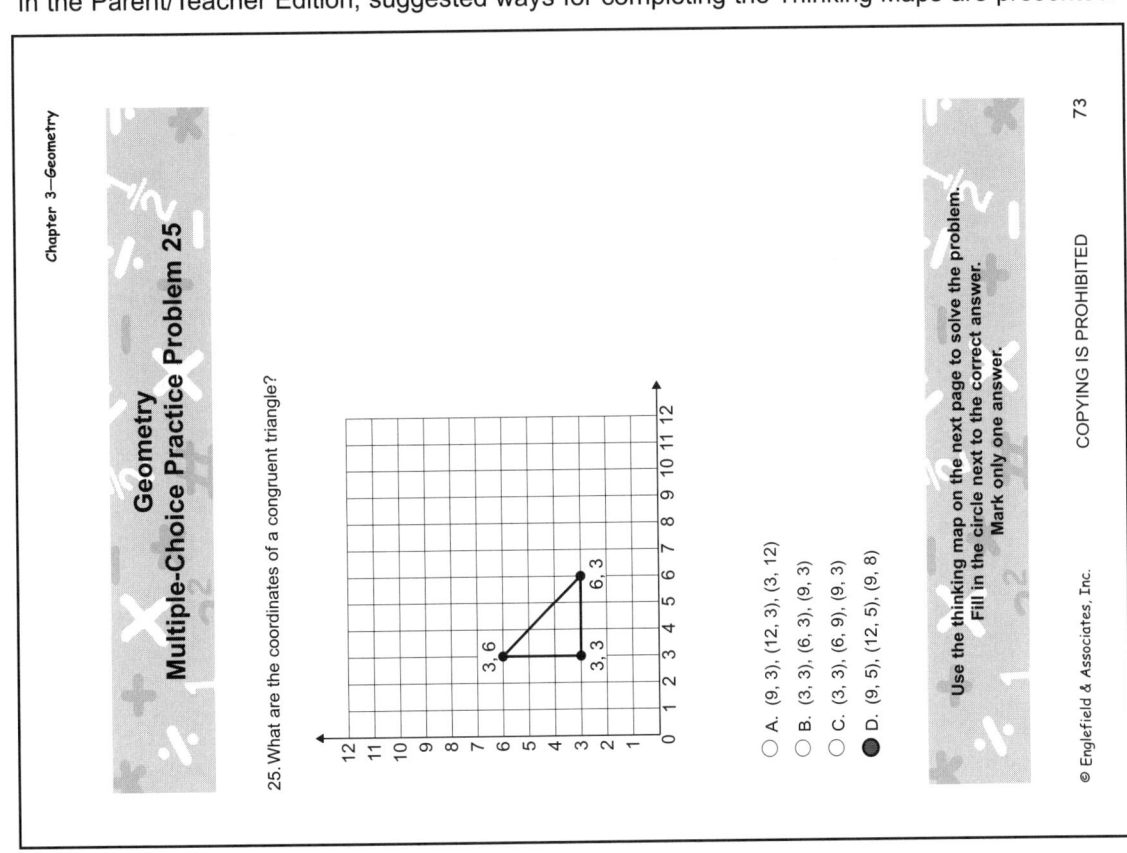

○ A. (9, 3), (12, 3), (3, 12)
○ B. (3, 3), (6, 3), (9, 3)
○ C. (3, 3), (6, 9), (9, 3)
● D. (9, 5), (12, 5), (9, 8)

Use the thinking map on the next page to solve the problem.
Fill in the circle next to the correct answer.
Mark only one answer.

Chapter 3—Geometry

Thinking Map

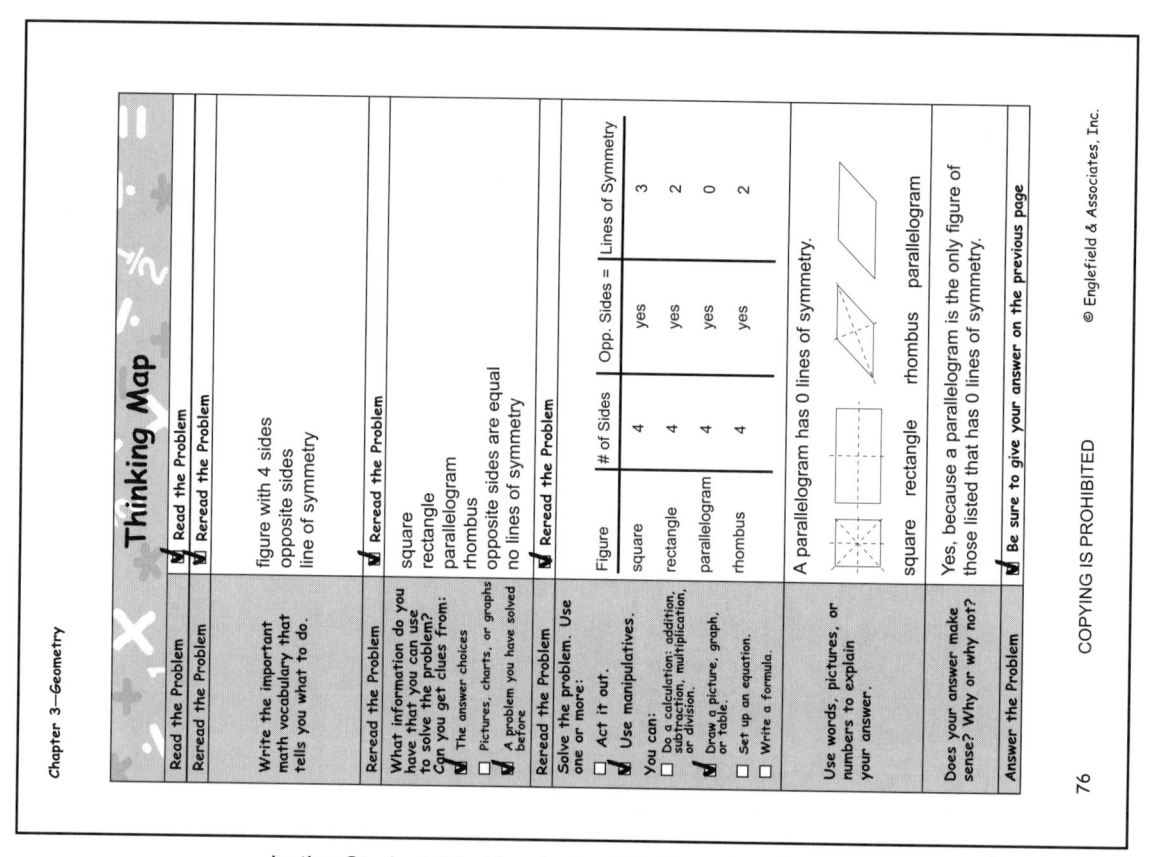

Read the Problem	✓ Read the Problem
Reread the Problem	✓ Reread the Problem
Write the important math vocabulary that tells you what to do.	figure with 4 sides opposite sides line of symmetry
Reread the Problem	✓ Reread the Problem
What information do you have that you can use to solve the problem? Can you get clues from: ☐ The answer choices ☐ Pictures, charts, or graphs ☑ A problem you have solved before	square rectangle parallelogram rhombus opposite sides are equal no lines of symmetry
Reread the Problem	✓ Reread the Problem
Solve the problem. Use one or more: ☐ Act it out. ☑ Use manipulatives. You can: ☐ Do a calculation: addition, subtraction, multiplication, or division. ☑ Draw a picture, graph, or table. ☐ Set up an equation. ☐ Write a formula.	Figure | # of Sides | Opp. Sides = | Lines of Symmetry square | 4 | yes | 3 rectangle | 4 | yes | 2 parallelogram | 4 | yes | 0 rhombus | 4 | yes | 2 A parallelogram has 0 lines of symmetry. square rectangle rhombus parallelogram
Use words, pictures, or numbers to explain your answer.	
Does your answer make sense? Why or why not?	Yes, because a parallelogram is the only figure of those listed that has 0 lines of symmetry.
Answer the Problem	✓ Be sure to give your answer on the previous page

In the Student Workbook, the Thinking Map is blank.
In the Parent/Teacher Edition, suggested ways for completing the Thinking Maps are presented.

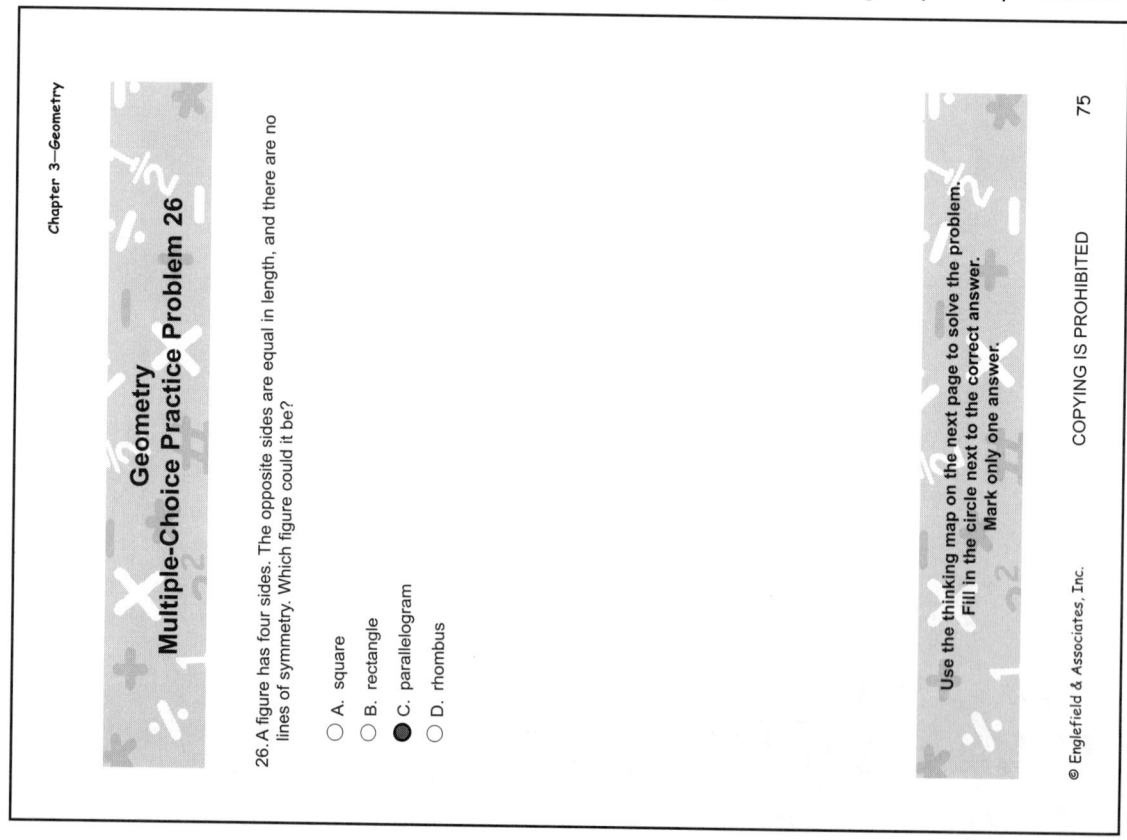

Geometry
Multiple-Choice Practice Problem 26

26. A figure has four sides. The opposite sides are equal in length, and there are no lines of symmetry. Which figure could it be?

○ A. square
○ B. rectangle
● C. parallelogram
○ D. rhombus

Use the thinking map on the next page to solve the problem.
Fill in the circle next to the correct answer.
Mark only one answer.

Thinking Map

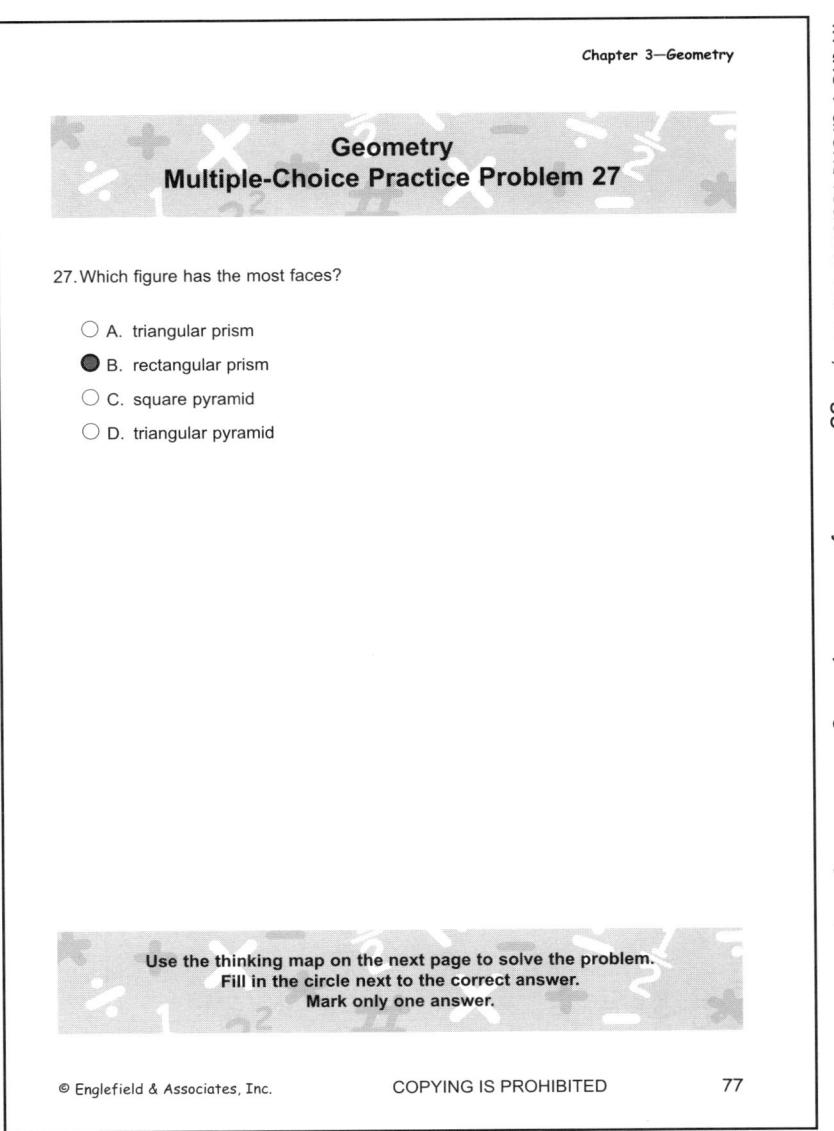

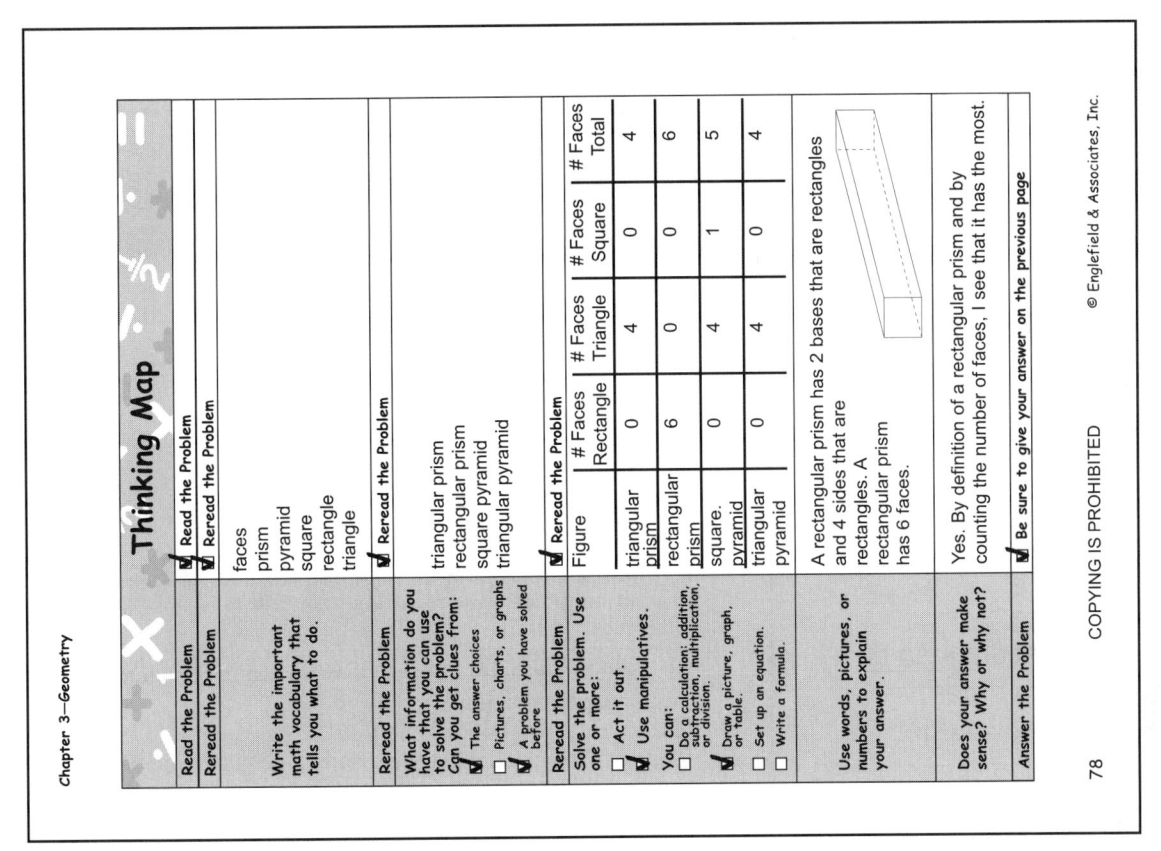

☑ Read the Problem	
☑ Reread the Problem	
Write the important math vocabulary that tells you what to do.	faces prism pyramid square rectangle triangle
☑ Reread the Problem	
What information do you have that you can use to solve the problem? Can you get clues from: ☐ The answer choices ☐ Pictures, charts, or graphs ☑ A problem you have solved before	triangular prism rectangular prism square pyramid triangular pyramid

☑ Reread the Problem

Figure	# Faces Rectangle	# Faces Triangle	# Faces Square	# Faces Total
triangular prism	0	4	0	4
rectangular prism	6	0	0	6
square pyramid	0	4	1	5
triangular pyramid	0	4	0	4

Solve the problem. Use one or more: ☐ Act it out. ☐ Use manipulatives. You can: ☐ Do a calculation: addition, subtraction, multiplication, or division. ☑ Draw a picture, graph, or table. ☐ Set up an equation. ☐ Write a formula.	A rectangular prism has 2 bases that are rectangles and 4 sides that are rectangles. A rectangular prism has 6 faces.
Use words, pictures, or numbers to explain your answer.	
Does your answer make sense? Why or why not?	Yes. By definition of a rectangular prism and by counting the number of faces, I see that it has the most.
Answer the Problem	☑ Be sure to give your answer on the previous page

In the Student Workbook, the Thinking Map is blank.
In the Parent/Teacher Edition, suggested ways for completing the Thinking Maps are presented.

Geometry
Multiple-Choice Practice Problem 27

27. Which figure has the most faces?

○ A. triangular prism
● B. rectangular prism
○ C. square pyramid
○ D. triangular pyramid

Use the thinking map on the next page to solve the problem.
Fill in the circle next to the correct answer.
Mark only one answer.

Chapter 3—Geometry

Model Problem 8:
Geometry
Short-Answer

1. Prior Knowledge

Determine the student's prior knowledge about the strand, content of the problem, and/or mathematical process of the problem.

These questions can be used to help determine students' knowledge:

- What do you know about quadrilaterals?
- When do you use them?

Additional prompts could be provided by using the following topics:

- mosaic art
- flooring tiles
- carpeting
- stained glass windows
- jewelry
- medical imagery
- clothing

**Model Problem 8:
Geometry
Short-Answer**

8. Is it possible to draw two quadrilaterals in which all pairs of corresponding sides are congruent in length, but the figures are not congruent? Name the two different figures.

> 1 point: The student attempts to draw two quadrilaterals with congruent sides.
>
> 1 point: The student identifies and names each figure.
>
> Answers may vary; any two quadrilaterals that fit the criteria are acceptable.

Use the thinking map on the next page to solve the problem.
Write your answer in the box.

2. Model

Model the use of the mathematical thinking map through class discussion with the thinking process modeled aloud as the thinking map is completed.

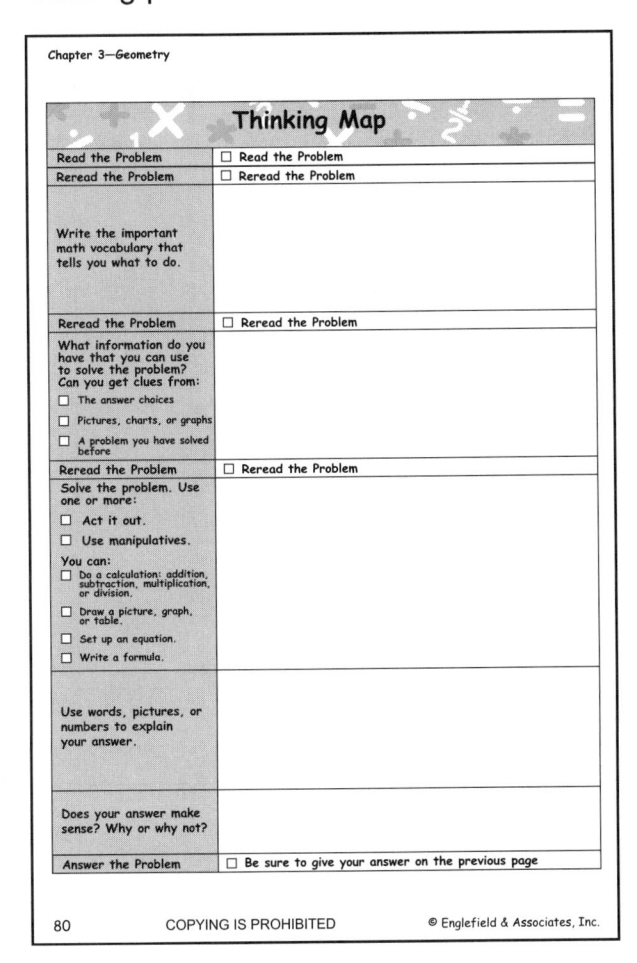

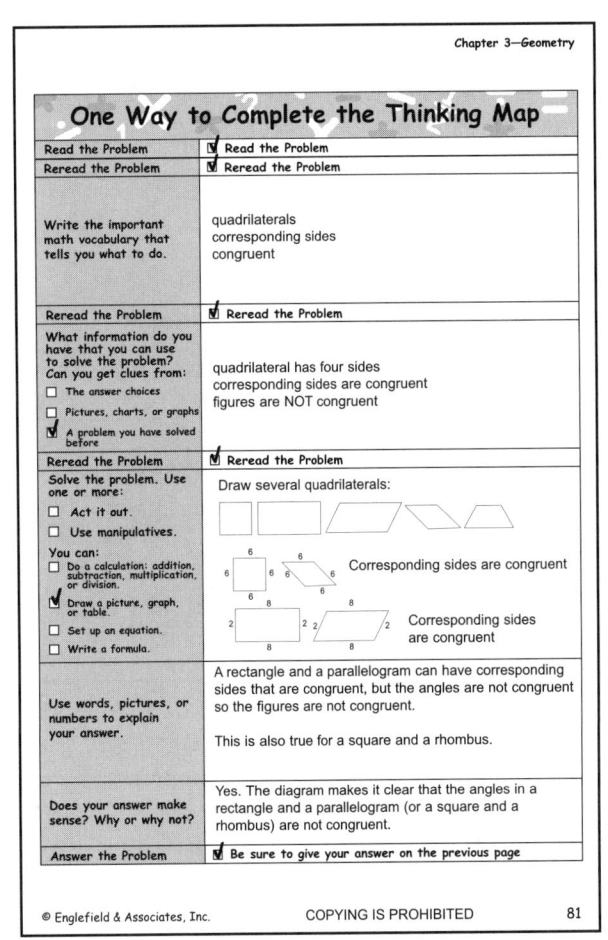

Thinking Map	
Read the Problem	☐ Read the Problem
Reread the Problem	☐ Reread the Problem
Write the important math vocabulary that tells you what to do.	
Reread the Problem	☐ Reread the Problem
What information do you have that you can use to solve the problem? Can you get clues from: ☐ The answer choices ☐ Pictures, charts, or graphs ☐ A problem you have solved before	
Reread the Problem	☐ Reread the Problem
Solve the problem. Use one or more: ☐ Act it out. ☐ Use manipulatives. You can: ☐ Do a calculation: addition, subtraction, multiplication, or division. ☐ Draw a picture, graph, or table. ☐ Set up an equation. ☐ Write a formula.	
Use words, pictures, or numbers to explain your answer.	
Does your answer make sense? Why or why not?	
Answer the Problem	☐ Be sure to give your answer on the previous page

One Way to Complete the Thinking Map	
Read the Problem	☑ Read the Problem
Reread the Problem	☑ Reread the Problem
Write the important math vocabulary that tells you what to do.	quadrilaterals corresponding sides congruent
Reread the Problem	☑ Reread the Problem
What information do you have that you can use to solve the problem? Can you get clues from: ☐ The answer choices ☐ Pictures, charts, or graphs ☑ A problem you have solved before	quadrilateral has four sides corresponding sides are congruent figures are NOT congruent
Reread the Problem	☑ Reread the Problem
Solve the problem. Use one or more: ☐ Act it out. ☐ Use manipulatives. You can: ☐ Do a calculation: addition, subtraction, multiplication, or division. ☑ Draw a picture, graph, or table. ☐ Set up an equation. ☐ Write a formula.	Draw several quadrilaterals: Corresponding sides are congruent Corresponding sides are congruent
Use words, pictures, or numbers to explain your answer.	A rectangle and a parallelogram can have corresponding sides that are congruent, but the angles are not congruent so the figures are not congruent. This is also true for a square and a rhombus.
Does your answer make sense? Why or why not?	Yes. The diagram makes it clear that the angles in a rectangle and a parallelogram (or a square and a rhombus) are not congruent.
Answer the Problem	☑ Be sure to give your answer on the previous page

3. Guided Practice

Provide guided practice by presenting a problem for students to complete that matches the strand, content, or process you want to teach.

4. Independent Practice

Provide independent practice so students are provided with multiple-choice, short-answer, or extended-response problems.

Chapter 3—Geometry

Thinking Map

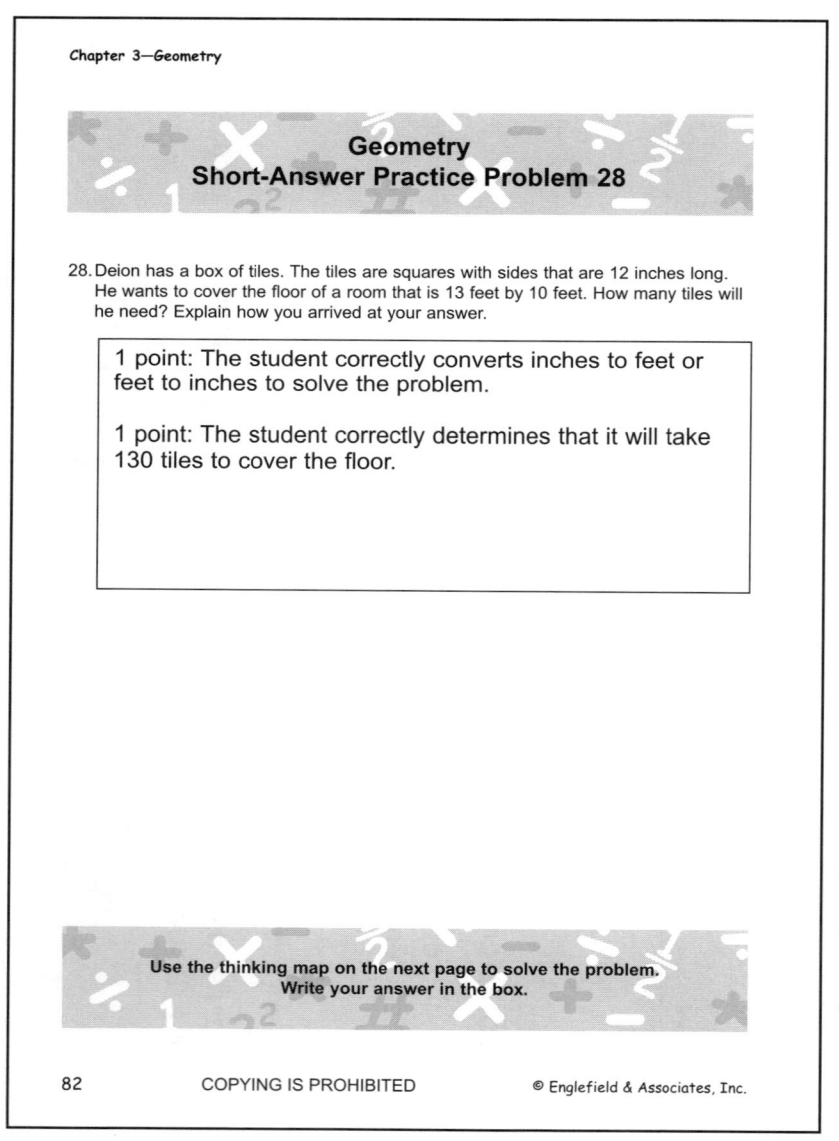

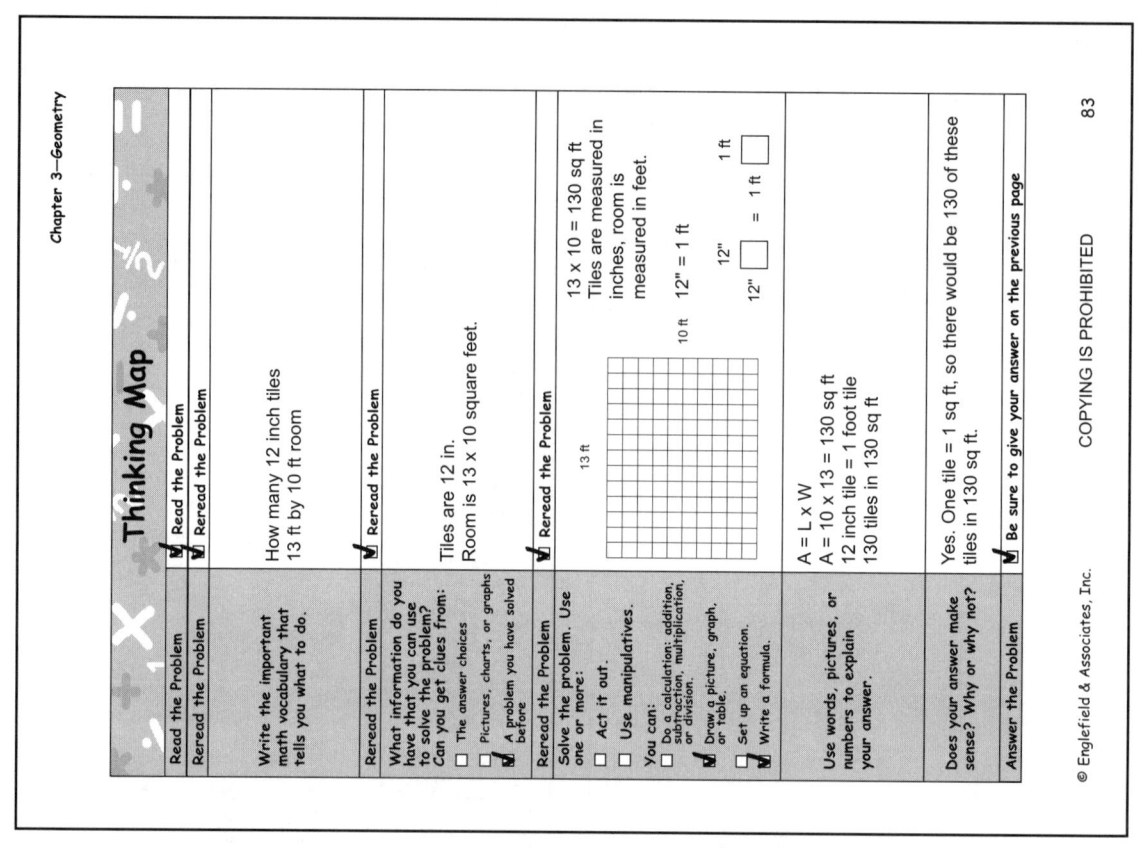

Read the Problem ✓	
Reread the Problem ✓	
Write the important math vocabulary that tells you what to do.	How many 12 inch tiles 13 ft by 10 ft room
Reread the Problem ✓	
What information do you have that you can use to solve the problem? Can you get clues from: ☐ The answer choices ☐ Pictures, charts, or graphs ✓ A problem you have solved before	Tiles are 12 in. Room is 13 x 10 square feet.
Reread the Problem ✓	
Solve the problem. Use one or more: ☐ Act it out. ☐ Use manipulatives. You can: ☐ Do a calculation: addition, subtraction, multiplication, or division. ✓ Draw a picture, graph, or table. ✓ Set up an equation. ✓ Write a formula.	13 x 10 = 130 sq ft Tiles are measured in inches, room is measured in feet. 10 ft [grid] 13 ft 12" = 1 ft 12" = 1 ft A = L x W A = 10 x 13 = 130 sq ft 12 inch tile = 1 foot tile 130 tiles in 130 sq ft
Use words, pictures, or numbers to explain your answer.	
Does your answer make sense? Why or why not?	Yes. One tile = 1 sq ft, so there would be 130 of these tiles in 130 sq ft.
Answer the Problem ✓	Be sure to give your answer on the previous page

In the Student Workbook, the Thinking Map is blank.
In the Parent/Teacher Edition, suggested ways for completing the Thinking Maps are presented.

Geometry
Short-Answer Practice Problem 28

28. Deion has a box of tiles. The tiles are squares with sides that are 12 inches long. He wants to cover the floor of a room that is 13 feet by 10 feet. How many tiles will he need? Explain how you arrived at your answer.

1 point: The student correctly converts inches to feet or feet to inches to solve the problem.

1 point: The student correctly determines that it will take 130 tiles to cover the floor.

Use the thinking map on the next page to solve the problem.
Write your answer in the box.

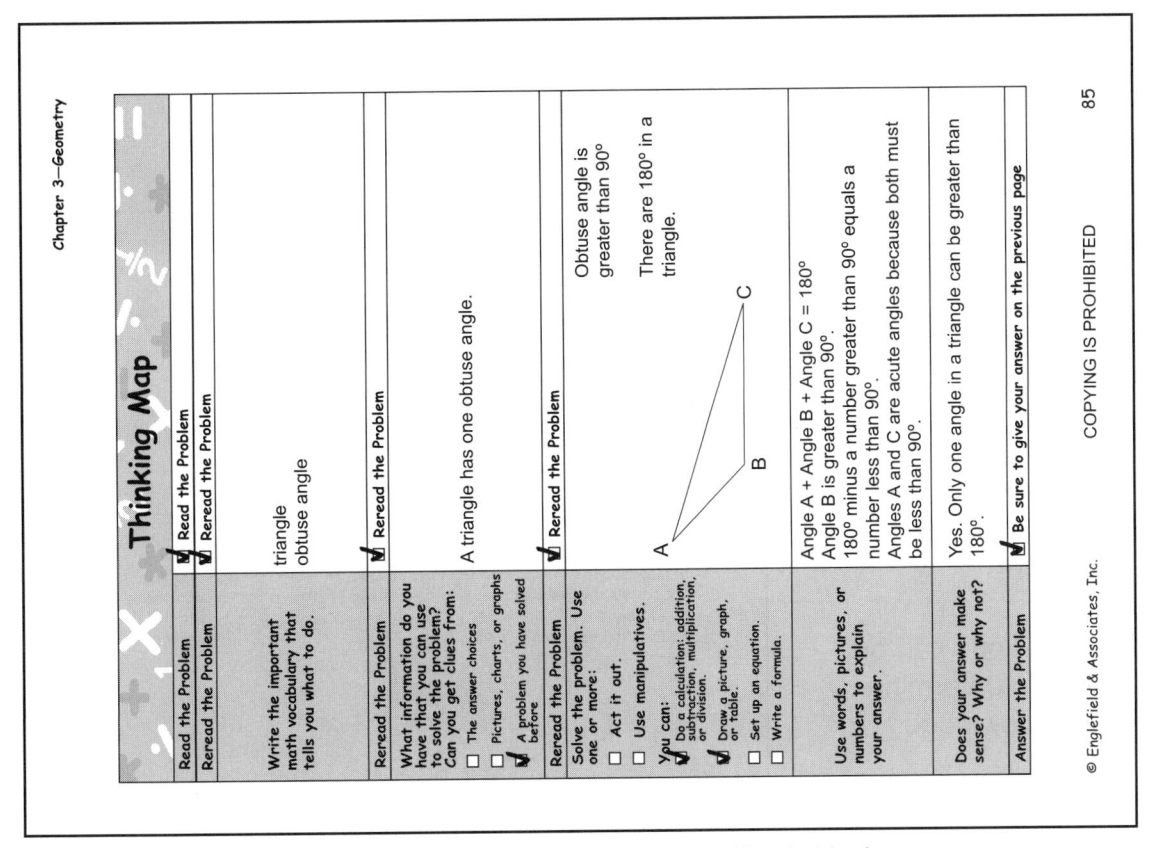

Geometry
Short-Answer Practice Problem 29

29. If one angle in a triangle is obtuse, what can you tell about the other two angles? Explain your answer using words or pictures.

> **1 point:** The student applies knowledge about obtuse and acute angles, and the student knows the sum of the angles in a triangle = 180°.
>
> **1 point:** The student determines that the other two angles must be acute.

Use the thinking map on the next page to solve the problem. Write your answer in the box.

Thinking Map

☑ Read the Problem	
☑ Reread the Problem	
Write the important math vocabulary that tells you what to do.	triangle obtuse angle
☑ Reread the Problem	
What information do you have that you can use to solve the problem? Can you get clues from: ☐ The answer choices ☐ Pictures, charts, or graphs ☑ A problem you have solved before	A triangle has one obtuse angle.
☑ Reread the Problem	
Solve the problem. Use one or more: ☐ Act it out. ☐ Use manipulatives. You can: ☐ Do a calculation: addition, subtraction, multiplication, or division. ☑ Draw a picture, graph, or table. ☐ Set up an equation. ☐ Write a formula. Use words, pictures, or numbers to explain your answer.	[Triangle ABC with obtuse angle at B] Obtuse angle is greater than 90°. There are 180° in a triangle. Angle A + Angle B + Angle C = 180°. Angle B is greater than 90°. 180° minus a number greater than 90° equals a number less than 90°. Angles A and C are acute angles because both must be less than 90°.
Does your answer make sense? Why or why not?	Yes. Only one angle in a triangle can be greater than 180°.
☑ Answer the Problem	Be sure to give your answer on the previous page

In the Student Workbook, the Thinking Map is blank.
In the Parent/Teacher Edition, suggested ways for completing the Thinking Maps are presented.

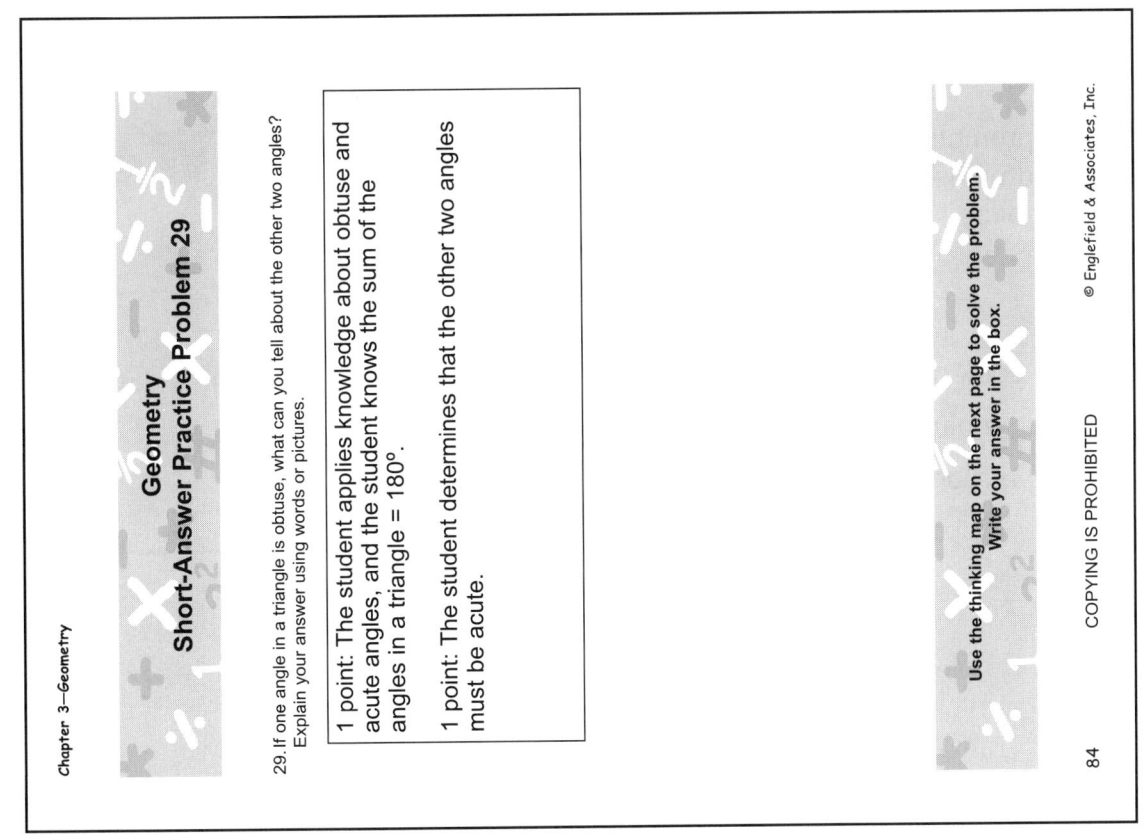

Chapter 3—Geometry

Model Problem 9: Geometry Extended-Response

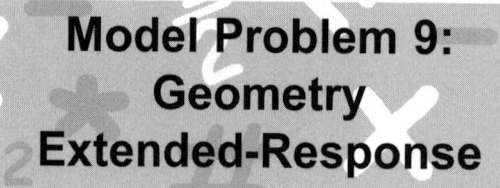

1. Prior Knowledge

Determine the student's prior knowledge about the strand, content of the problem, and/or mathematical process of the problem.

These questions can be used to help determine students' knowledge:

- What do you know about perimeter? Area?
- When do you use perimeter?
- When do you use area?

Additional prompts could be provided by using the following topics:

- fences
- carpet
- making book covers
- wrapping a gift
- painting a room

Model Problem 9: Geometry Extended-Response

9. Elsa is baking brownies, and the recipe calls for a 9" x 13" baking pan. Elsa does not have a pan that is this size. She has two of each of the following pans: an 8" x 8" square pan and an 8" round pie pan. What is the best combination of pans for Elsa to use and why?

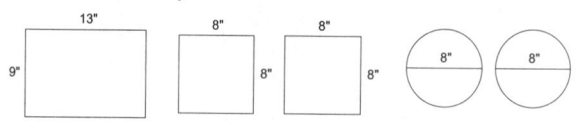

[13" x 9" rectangle] [8" x 8" square] [8" x 8" square] [8" circle] [8" circle]

1 point: The student demonstrates an understanding of area and how to compute area.

1 point: The student compares the areas of the two pans.

1 point: The student correctly determines that the 8" x 8" square pans would be the best to use.

1 point: The student explains the reason behind the answer. This problem can be solved without figuring out the area of the round pans.

Use the thinking map on the next page to solve the problem.
Write your answer in the box.

2. Model

Model the use of the mathematical thinking map through class discussion with the thinking process modeled aloud as the thinking map is completed.

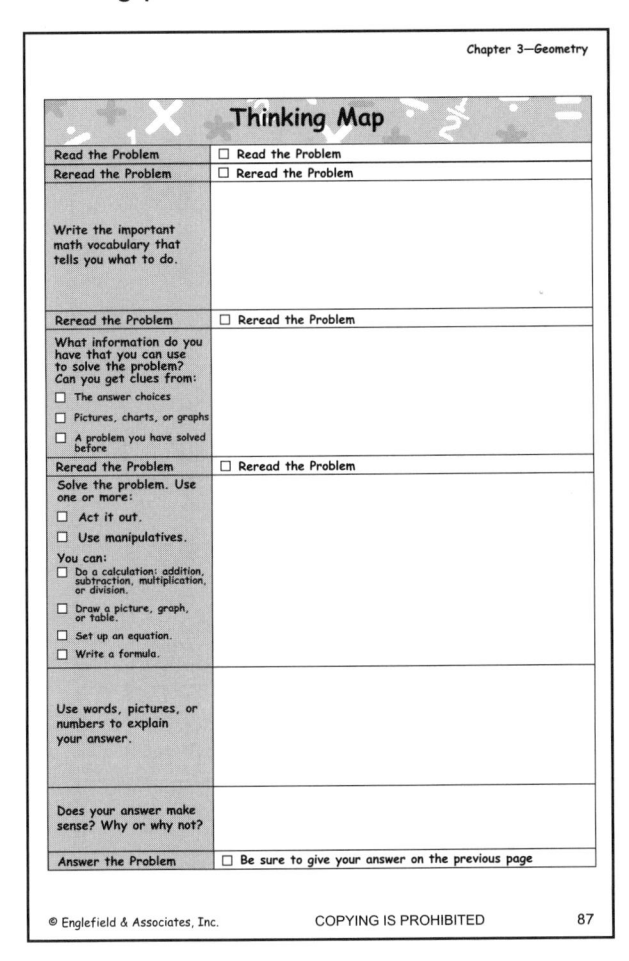

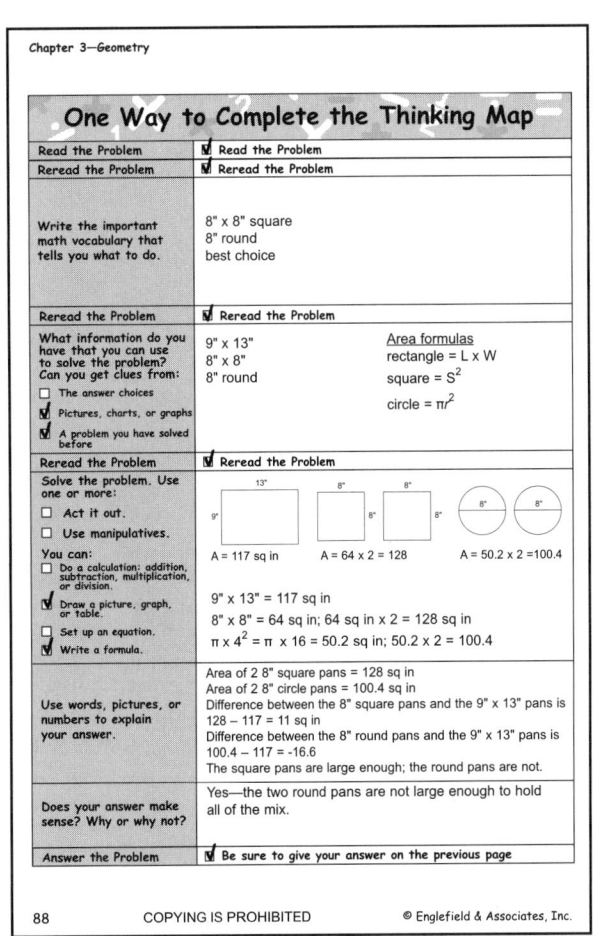

Thinking Map

Read the Problem	☐ Read the Problem
Reread the Problem	☐ Reread the Problem
Write the important math vocabulary that tells you what to do.	
Reread the Problem	☐ Reread the Problem
What information do you have that you can use to solve the problem? Can you get clues from: ☐ The answer choices ☐ Pictures, charts, or graphs ☐ A problem you have solved before	
Reread the Problem	☐ Reread the Problem
Solve the problem. Use one or more: ☐ Act it out. ☐ Use manipulatives. You can: ☐ Do a calculation: addition, subtraction, multiplication, or division. ☐ Draw a picture, graph, or table. ☐ Set up an equation. ☐ Write a formula.	
Use words, pictures, or numbers to explain your answer.	
Does your answer make sense? Why or why not?	
Answer the Problem	☐ Be sure to give your answer on the previous page

One Way to Complete the Thinking Map

Read the Problem	☑ Read the Problem
Reread the Problem	☑ Reread the Problem
Write the important math vocabulary that tells you what to do.	8" x 8" square 8" round best choice
Reread the Problem	☑ Reread the Problem
What information do you have that you can use to solve the problem? Can you get clues from: ☐ The answer choices ☑ Pictures, charts, or graphs ☑ A problem you have solved before	9" x 13" 8" x 8" 8" round Area formulas rectangle = L x W square = S^2 circle = πr^2
Reread the Problem	☑ Reread the Problem
Solve the problem. Use one or more: ☐ Act it out. ☐ Use manipulatives. You can: ☐ Do a calculation: addition, subtraction, multiplication, or division. ☑ Draw a picture, graph, or table. ☐ Set up an equation. ☑ Write a formula.	A = 117 sq in A = 64 x 2 = 128 A = 50.2 x 2 = 100.4 9" x 13" = 117 sq in 8" x 8" = 64 sq in; 64 sq in x 2 = 128 sq in $\pi \times 4^2 = \pi \times 16 = 50.2$ sq in; 50.2 x 2 = 100.4
Use words, pictures, or numbers to explain your answer.	Area of 2 8" square pans = 128 sq in Area of 2 8" circle pans = 100.4 sq in Difference between the 8" square pans and the 9" x 13" pans is 128 – 117 = 11 sq in Difference between the 8" round pans and the 9" x 13" pans is 100.4 – 117 = -16.6 The square pans are large enough; the round pans are not.
Does your answer make sense? Why or why not?	Yes—the two round pans are not large enough to hold all of the mix.
Answer the Problem	☑ Be sure to give your answer on the previous page

3. Guided Practice

Provide guided practice by presenting a problem for students to complete that matches the strand, content, or process you want to teach.

4. Independent Practice

Provide independent practice so students are provided with multiple-choice, short-answer, or extended-response problems.

Chapter 3—Geometry

Thinking Map

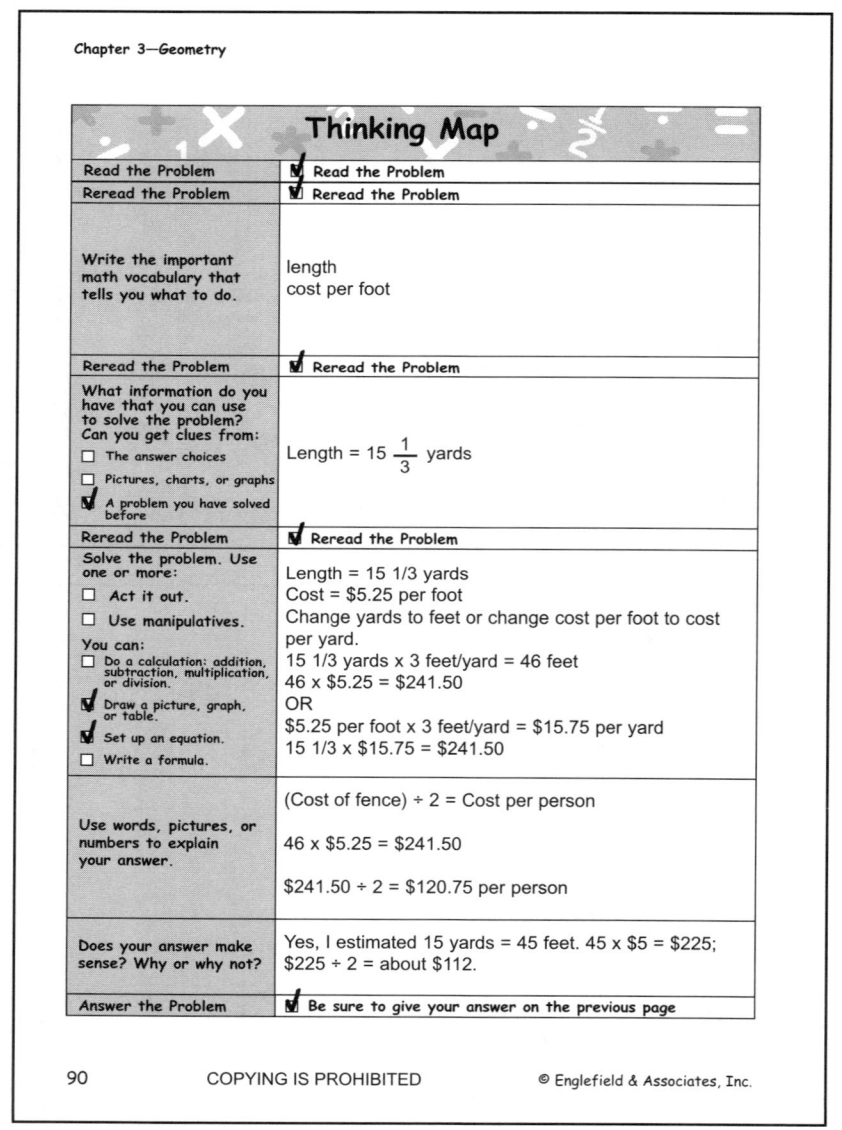

Read the Problem	✓ Read the Problem
Reread the Problem	✓ Reread the Problem
Write the important math vocabulary that tells you what to do.	length cost per foot
Reread the Problem	✓ Reread the Problem
What information do you have that you can use to solve the problem? Can you get clues from: ☐ The answer choices ✓ A problem you have solved before ☐ Pictures, charts, or graphs	Length = $15\frac{1}{3}$ yards
Reread the Problem	✓ Reread the Problem
Solve the problem. Use one or more: ☐ Act it out. ☐ Use manipulatives. You can: ☐ Do a calculation: addition, subtraction, multiplication, or division. ✓ Draw a picture, graph, or table. ✓ Set up an equation. ☐ Write a formula.	Length = 15 1/3 yards Cost = $5.25 per foot Change yards to feet or change cost per foot to cost per yard. 15 1/3 yards × 3 feet/yard = 46 feet 46 × $5.25 = $241.50 OR $5.25 per foot × 3 feet/yard = $15.75 per yard 15 1/3 × $15.75 = $241.50
Use words, pictures, or numbers to explain your answer.	(Cost of fence) ÷ 2 = Cost per person 46 × $5.25 = $241.50 $241.50 ÷ 2 = $120.75 per person
Does your answer make sense? Why or why not?	Yes, I estimated 15 yards = 45 feet. 45 × $5 = $225; $225 ÷ 2 = about $112.
Answer the Problem	✓ Be sure to give your answer on the previous page

---

In the Student Workbook, the Thinking Map is blank.
In the Parent/Teacher Edition, suggested ways for completing the Thinking Maps are presented.

Geometry
Extended-Response Practice Problem 30

30. Mr. Timm and his neighbor decided to fence their yards and to share the cost of the fence that borders both yards. Mr. Timm measured the length of that section of fence as $15\frac{1}{3}$ yards. The cost of the fencing is $5.25 per foot. How much will each pay for that length of fence? Explain how you arrived at your answer.

1 point: The student correctly converts yards to feet to find the length of the section in feet.

1 point: The student recognizes that the total cost of the section of fence is found by multiplying the length (in feet) by the cost per foot.

1 point: The student correctly finds the cost of the length of fence.

1 point: The student correctly calculates how much each of the neighbors will pay for the fence.

Use the thinking map on the next page to solve the problem. Write your answer in the box.

Literature Connections for Geometry

Arrow to the Sun: A Pueblo Indian Tale
Gerald McDermott
© 1977

The Patchwork Quilt
Valerie Flournoy and Jerry Pinkney
© 1985

Cubes, Cones, Cylinders, & Spheres
Tana Hoban
© 2000

The Keeping Quilt
Patricia Polacco
© 2001

A Cloak for the Dreamers
Aileen Freidman
© 1995

The Greedy Triangle
Marilyn Burns
© 1995

Kitten Castle
Mel Friedman and
Ellen Weiss
© 2001

The Seasons Sewn: A Year in Patchwork
Ann Whitford Paul
© 1996

Shape Up!
David A. Adler
© 2000

Angles Are Easy as Pie
Robert Froman
© 1976

Circles: Fun Ideas for Getting A-Round in Math
Catherine Sheldrick Ross
© 1993

Mirror Magic
Seymour Simon
© 1980

Literature Connections for Geometry (Continued)

Pyramid
David MacAulay
© 1982

Right Angles: Paper-Folding Geometry
Jo McKeeby Phillips
© 1972

Sir Cumference and the First Round Table
Cindy Neuschwander
© 1997

Sir Cumference and the Sword in the Cone
Cindy Neuschwander
© 2003

Sir Cumference and the Great Knight of Angleland
Cindy Neuschwander
© 2001

Stanley, Flat Again!
Jeff Brown
© 2003

The Boy with Square Eyes
Juliet Snape
© 1988

The Tangram Magician
Lisa Campbell Ernst
© 1990

Grandfather Tang's Story
Ann Tompert
© 1990

Sweet Clara and the Freedom Quilt
Deborah Hopkinson
© 1993

Conned Again, Watson! Cautionary Tales of Logic, Math, and Probability
Colin Bruce
© 2001

A Cloak for the Dreamer
Aileen Friedman
© 1995

Chapter 4
Algebra

The purposes of this chapter include

1. Providing a definition of Algebra and what it means for fourth-grade students.
2. Offering teaching tips that reinforce Algebra concepts.
3. Offering multiple-choice, short-answer, and extended-response questions dealing with Algebra.
4. Providing ideas for connecting Algebra concepts and processes with other areas of the curriculum.

The following teaching tools are provided for extending students' thinking with Algebra:

- Model lesson for Algebra multiple-choice problems
- Seven Algebra multiple-choice practice problems
- Model lesson for Algebra short-answer problems
- Two Algebra short-answer practice problems
- Model lesson for Algebra extended-response problems
- One Algebra extended-response practice problem

What is Algebra?

In grade 4, students examine numerical and geometric patterns as they express the patterns through words, symbols, or numbers. Through observations of the changes in patterns, students formulate rules or generalizations. The concept of variables is explored using a blank box, a letter, or a symbol to represent the unknown quantity. Eventually, students will use equations to represent different patterns of change.

What does Algebra look like?

- Students describe patterns and represent them using tables or symbols.
- Students make predictions and draw conclusions as they observe models, tables, charts, or graphs.
- Students will work with variables represented by boxes, letters, or symbols.
- Students will develop strategies to solve multiplication problems.
- Students will identify patterns, often using real-world data to realize that predictions will not always match outcomes.
- Students will study different patterns of change as preparation for more advanced mathematical concepts.
- Students will describe patterns of change that they see in geometric shapes and will express the changes numerically.

Vocabulary related to Algebra:

Arithmetic sequence: An ordered set of numbers, shapes, or other mathematical objects arranged according to a rule.

Commutative property: A property that applies to adding and multiplying numbers. Numbers can be added together or multiplied together in any order. For example, 7 + 3 = 10 is the same as 3 + 7 = 10; 6 x 2 = 12 is the same as 2 x 6 = 12.

Equation: Statement that shows two mathematical expressions that are equal to each other.

Formula: A mathematical rule written using symbols, usually as an equation describing a certain relationship between quantities.

Functions: Relationships between two values. The second value is a function of the first one. For example, $3(a) = b$. The first variable is a; the second is b. If you plug in a number for variable a, it affects variable b: $3(2) = 6$.

Inequalities: Things not equal in amount or value, represented by "≠"; 4 ≠ 6.

Number sentence: A mathematical sentence written in numerals or mathematical symbols.

Pattern: A repeated design or recurring sequence.

Quantitative change: A change that can be counted or measured.

Rule: A standard method or procedure for solving a class of problems.

Unknown quantity: An amount or number of something that is not known.

Variable: A changing quantity that might have a range of possible values, usually a letter in an algebraic equation or expression.

Chapter 4—Algebra

Teaching Tips

- Teachers can use graphed squares for students to express patterns in mathematical sentences.
- Teachers can ask students to use words, numbers, or symbols to make generalizations based on patterns that they observe in geometric shapes.
- Teachers can ask questions to illustrate the properties of numbers. Is 6 x 3 equal to 3 x 6? How do you know? Is 4 x 24 the same as 4 x 20 + 4 x 4?
- Teachers can ask students to observe temperature, plant growth, or any changing event to mathematically describe change.

For Future Reference

Note different strategies students used to solve the problems.

Chapter 4—Algebra

Model Problem 10: Algebra Multiple-Choice

1. Prior Knowledge

Determine the student's prior knowledge about the strand, content of the problem, and/or mathematical process of the problem.

These questions can be used to help determine students' knowledge:

- When you are having difficulty solving a problem, what are the ways you can simplify the problem?

- When are some times that you have successfully simplified a problem?

Additional prompts could be provided by using the following topics:

- recipes
- solving for an unknown (algebraic equations)
- measurements for building or sewing objects
- working with statistics in social studies or science

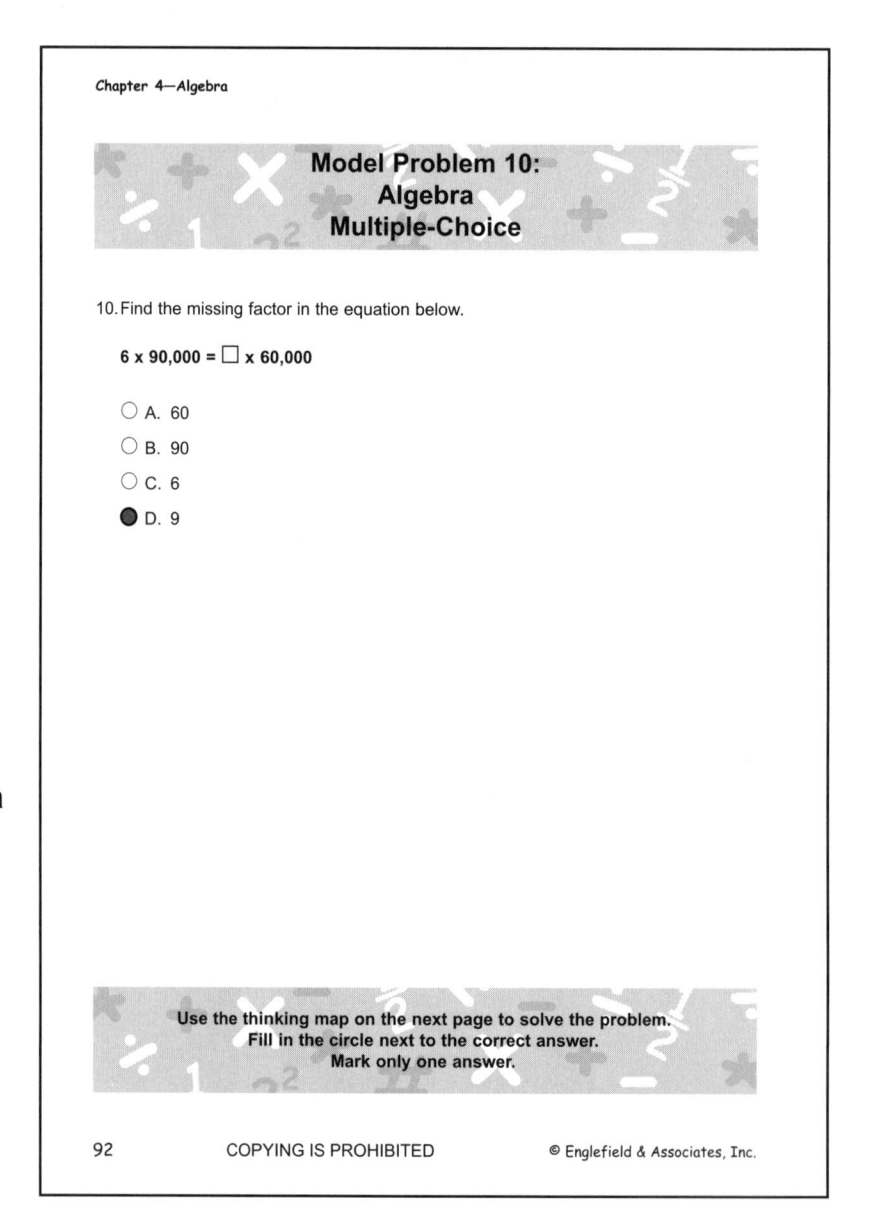

Chapter 4—Algebra

Model Problem 10: Algebra Multiple-Choice

10. Find the missing factor in the equation below.

    $6 \times 90{,}000 = \square \times 60{,}000$

    ○ A. 60
    ○ B. 90
    ○ C. 6
    ● D. 9

Use the thinking map on the next page to solve the problem.
Fill in the circle next to the correct answer.
Mark only one answer.

2. Model

Model the use of the mathematical thinking map through class discussion with the thinking process modeled aloud as the thinking map is completed.

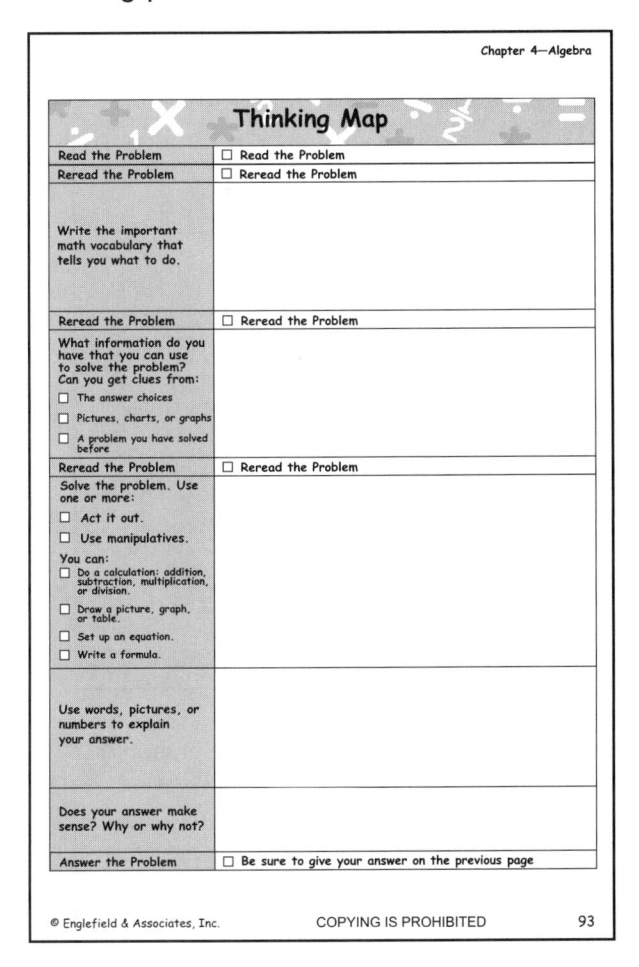

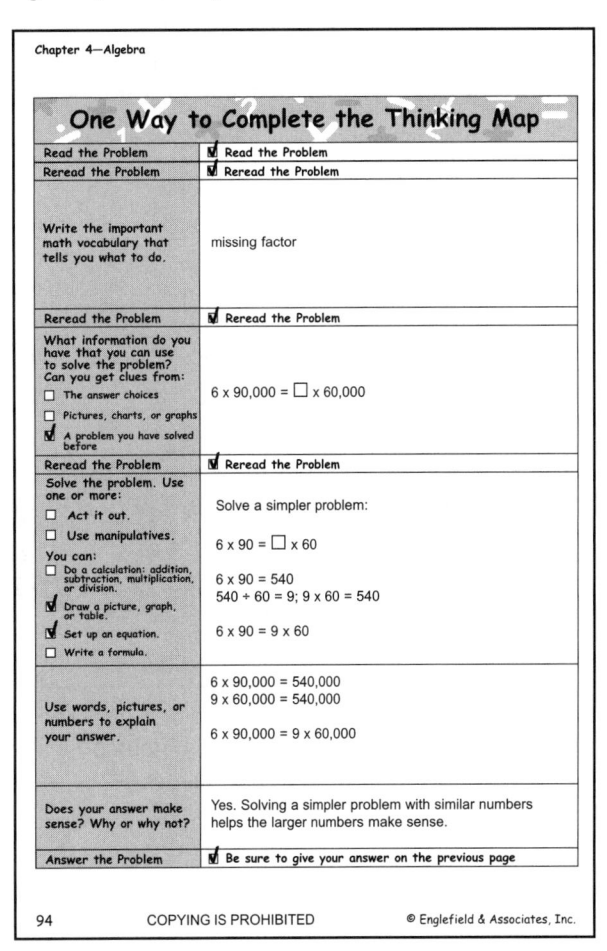

Thinking Map	
Read the Problem	☐ Read the Problem
Reread the Problem	☐ Reread the Problem
Write the important math vocabulary that tells you what to do.	
Reread the Problem	☐ Reread the Problem
What information do you have that you can use to solve the problem? Can you get clues from: ☐ The answer choices ☐ Pictures, charts, or graphs ☐ A problem you have solved before	
Reread the Problem	☐ Reread the Problem
Solve the problem. Use one or more: ☐ Act it out. ☐ Use manipulatives. You can: ☐ Do a calculation: addition, subtraction, multiplication, or division. ☐ Draw a picture, graph, or table. ☐ Set up an equation. ☐ Write a formula.	
Use words, pictures, or numbers to explain your answer.	
Does your answer make sense? Why or why not?	
Answer the Problem	☐ Be sure to give your answer on the previous page

One Way to Complete the Thinking Map	
Read the Problem	☑ Read the Problem
Reread the Problem	☑ Reread the Problem
Write the important math vocabulary that tells you what to do.	missing factor
Reread the Problem	☑ Reread the Problem
What information do you have that you can use to solve the problem? Can you get clues from: ☐ The answer choices ☐ Pictures, charts, or graphs ☑ A problem you have solved before	6 × 90,000 = ☐ × 60,000
Reread the Problem	☑ Reread the Problem
Solve the problem. Use one or more: ☐ Act it out. ☐ Use manipulatives. You can: ☐ Do a calculation: addition, subtraction, multiplication, or division. ☑ Draw a picture, graph, or table. ☑ Set up an equation. ☐ Write a formula.	Solve a simpler problem: 6 × 90 = ☐ × 60 6 × 90 = 540 540 ÷ 60 = 9; 9 × 60 = 540 6 × 90 = 9 × 60
Use words, pictures, or numbers to explain your answer.	6 × 90,000 = 540,000 9 × 60,000 = 540,000 6 × 90,000 = 9 × 60,000
Does your answer make sense? Why or why not?	Yes. Solving a simpler problem with similar numbers helps the larger numbers make sense.
Answer the Problem	☑ Be sure to give your answer on the previous page

3. Guided Practice

Provide guided practice by presenting a problem for students to complete that matches the strand, content, or process you want to teach.

4. Independent Practice

Provide independent practice so students are provided with multiple-choice, short-answer, or extended-response problems.

Chapter 4—Algebra

Thinking Map

☑ **Read the Problem**	
☑ **Reread the Problem**	
Write the important math vocabulary that tells you what to do.	equation function machine
☑ **Reread the Problem**	
What information do you have that you can use to solve the problem? Can you get clues from: ☑ The answer choices ☐ Pictures, charts, or graphs ☐ A problem you have solved before	There are four sets of numbers to help you find the function: 2 | 5 3 | 7 4 | 9 5 | 11
☑ **Reread the Problem**	
Solve the problem. Use one or more: ☐ Act it out. ☑ Use manipulatives. You can: ☑ Do a calculation: addition, subtraction, multiplication, or division. ☐ Draw a picture, graph, or table. ☑ Set up an equation. ☐ Write a formula.	One way to solve the problem is to use counters and to use the guess-and-check method to discover the relationship between A and B. $2 + 2 + 1 = 5$ $3 + 3 + 1 = 7$ $4 + 4 + 1 = 9$ $5 + 5 + 1 = 11$ $B = (A \times 2) + 1$ Another way to solve the problem is to look at the answer choices and solve the equation by substituting values of A and B.
Use words, pictures, or numbers to explain your answer.	A. $5 = 2 + 3; 7 \neq 3 + 3$ B. $2 \neq 5 + 2$ C. $5 = (2 \times 2) + 1; 7 = (3 \times 2) + 1; 9 = (4 \times 2) + 1;$ $11 = (5 \times 2) + 1$ D. $5 - 2 = 3; 7 - 3 \neq 3$
Does your answer make sense? Why or why not?	Yes, the answer makes sense because all of the numbers fit correctly into equation C.
☑ **Answer the Problem**	Be sure to give your answer on the previous page

96

In the Student Workbook, the Thinking Map is blank.
In the Parent/Teacher Edition, suggested ways for completing the Thinking Maps are presented.

Algebra
Multiple-Choice Practice Problem 31

31. Which of the following shows an equation for the function machine below?

IN	OUT
A	B
2	5
3	7
4	9
5	11

○ A. $B = A + 3$
○ B. $A = B + 2$
● C. $B = (A \times 2) + 1$
○ D. $B - A = 3$

Use the thinking map on the next page to solve the problem.
Fill in the circle next to the correct answer.
Mark only one answer.

95

Chapter 4—Algebra

Thinking Map

Read the Problem	☑ Read the Problem
Reread the Problem	☑ Reread the Problem
Write the important math vocabulary that tells you what to do.	pattern
Reread the Problem	☑ Reread the Problem
What information do you have that you can use to solve the problem? Can you get clues from: ☐ The answer choices ☑ Pictures, charts, or graphs ☐ A problem you have solved before	Study the number sequence and look for patterns. 9 10 11 12
Reread the Problem	☑ Reread the Problem
Solve the problem. Use one or more: ☐ Act it out. ☐ Use manipulatives. You can: ☑ Do a calculation: addition, subtraction, multiplication, or division. ☑ Draw a picture, graph, or table. ☐ Set up an equation. ☐ Write a formula.	Make a table: | From | To | Operation | |---|---|---| | 3 | 6 | +3 | | 6 | 4 | –2 | | 4 | 7 | +3 | | 7 | 5 | –2 | | 5 | 8 | +3 | (continued) | From | To | Operation | |---|---|---| | 8 | 6 | –2 | | 6 | 9 | +3 | | 9 | 7 | –2 | | 7 | ? | +3 | The table makes it easier to see the pattern, which is +3, –2, +3, –2, etc. The last number given is 7, and the last operation performed is –2. This means the next number is 7 + 3 = 10.
Use words, pictures, or numbers to explain your answer.	
Does your answer make sense? Why or why not?	Yes. If you continue the table, you would come up with 9 as the next number, then 7, then 10, which is the answer.
Answer the Problem	☑ Be sure to give your answer on the previous page

In the Student Workbook, the Thinking Map is blank.
In the Parent/Teacher Edition, suggested ways for completing the Thinking Maps are presented.

Algebra
Multiple-Choice Practice Problem 32

32. What is the next number in the pattern shown below?

3, 6, 4, 7, 5, 8, 6, 9, 7, . . .

○ A. 9
● B. 10
○ C. 11
○ D. 5

Use the thinking map on the next page to solve the problem.
Fill in the circle next to the correct answer.
Mark only one answer.

Chapter 4—Algebra

Thinking Map

Read the Problem	✓ Read the Problem ✓ Reread the Problem
Write the important math vocabulary that tells you what to do.	percent (%) discount
Reread the Problem	✓ Reread the Problem
What information do you have that you can use to solve the problem? Can you get clues from: ✓ The answer choices ☐ Pictures, charts, or graphs ✓ A problem you have solved before	15% week 1 20% week 2 25% week 3, etc. Original cost of toy = $50 When will toy cost $25?
Reread the Problem	✓ Reread the Problem
Solve the problem. Use one or more: ☐ Act it out. ☐ Use manipulatives. You can: ☐ Do a calculation: addition, subtraction, multiplication, or division. ✓ Draw a picture, graph, or table. ☐ Set up an equation. ☐ Write a formula.	$25 = 1/2 of $50 = 50% of $50 Make a chart to find out which week will have a 50% discount. | Week # | 1 | 2 | 3 | 4 | 5 | 6 | 7 | 8 | | % Off | 15 | 20 | 25 | 30 | 35 | 40 | 45 | 50 |
Use words, pictures, or numbers to explain your answer.	Since $25 is half of $50, the problem can be solved by finding out which week has a 50% discount. You can make a chart to find the answer, or you can count how many times you add 5 to 15 to get to 50, since the discount increases by 5% each week.
Does your answer make sense? Why or why not?	Yes, you can easily check by counting by 5s from 15 to 50.
Answer the Problem	✓ Be sure to give your answer on the previous page

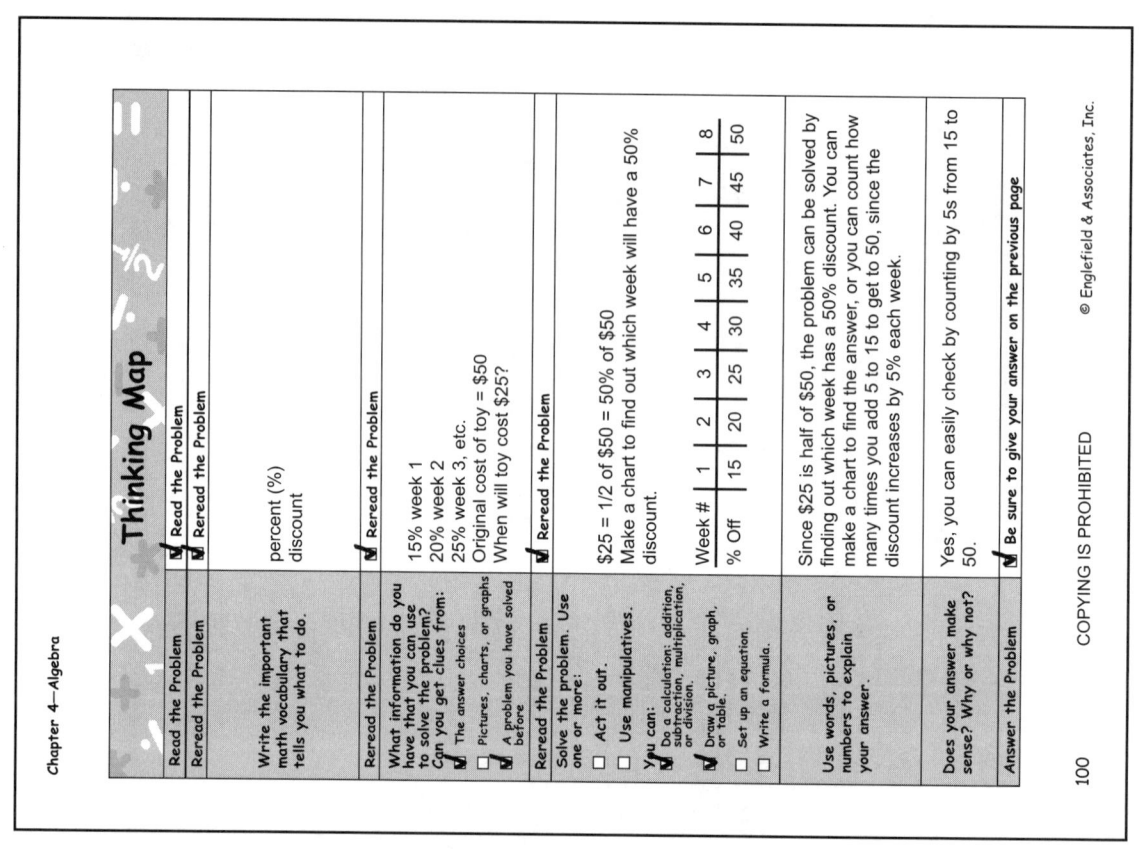

In the Student Workbook, the Thinking Map is blank.
In the Parent/Teacher Edition, suggested ways for completing the Thinking Maps are presented.

Algebra
Multiple-Choice Practice Problem 33

33. A toy store is going out of business and is putting all of the toys on sale until everything is sold. Toys will be on sale for 15% off the first week, 20% off the second week, 25% off the third week, etc. During what week will a game that is originally $50.00 cost $25.00?

○ A. Week 4
○ B. Week 5
● C. Week 8
○ D. Week 9

Use the thinking map on the next page to solve the problem.
Fill in the circle next to the correct answer.
Mark only one answer.

Thinking Map

Read the Problem	✓ Read the Problem
Reread the Problem	✓ Reread the Problem
Write the important math vocabulary that tells you what to do.	algebraic expression
Reread the Problem	✓ Reread the Problem
What information do you have that you can use to solve the problem? Can you get clues from: ✓ The answer choices ☐ Pictures, charts, or graphs ✓ A problem you have solved before	T = Tim's age D = Darcy's age
Reread the Problem	✓ Reread the Problem
Solve the problem. Use one or more: ☐ Act it out. ☐ Use manipulatives. You can: ✓ Do a calculation: addition, subtraction, multiplication, or division. ☐ Draw a picture, graph, or table. ✓ Set up an equation. ☐ Write a formula.	Tim's age (T) is 5 times Darcy's (D) age. So, Tim's age is equal to Darcy's age times 5: $T = 5 \times D$
Use words, pictures, or numbers to explain your answer.	The equation $T = 5 \times D$ can be written in words as "Tim's age is equal to 5 times Darcy's age" or "Tim (T) equals 5 times Darcy ($5 \times D$)."
Does your answer make sense? Why or why not?	Yes. You can plug in ages to check: if Darcy is 10 years old, Tim is 5 x 10 = 50 years old.
Answer the Problem	✓ Be sure to give your answer on the previous page

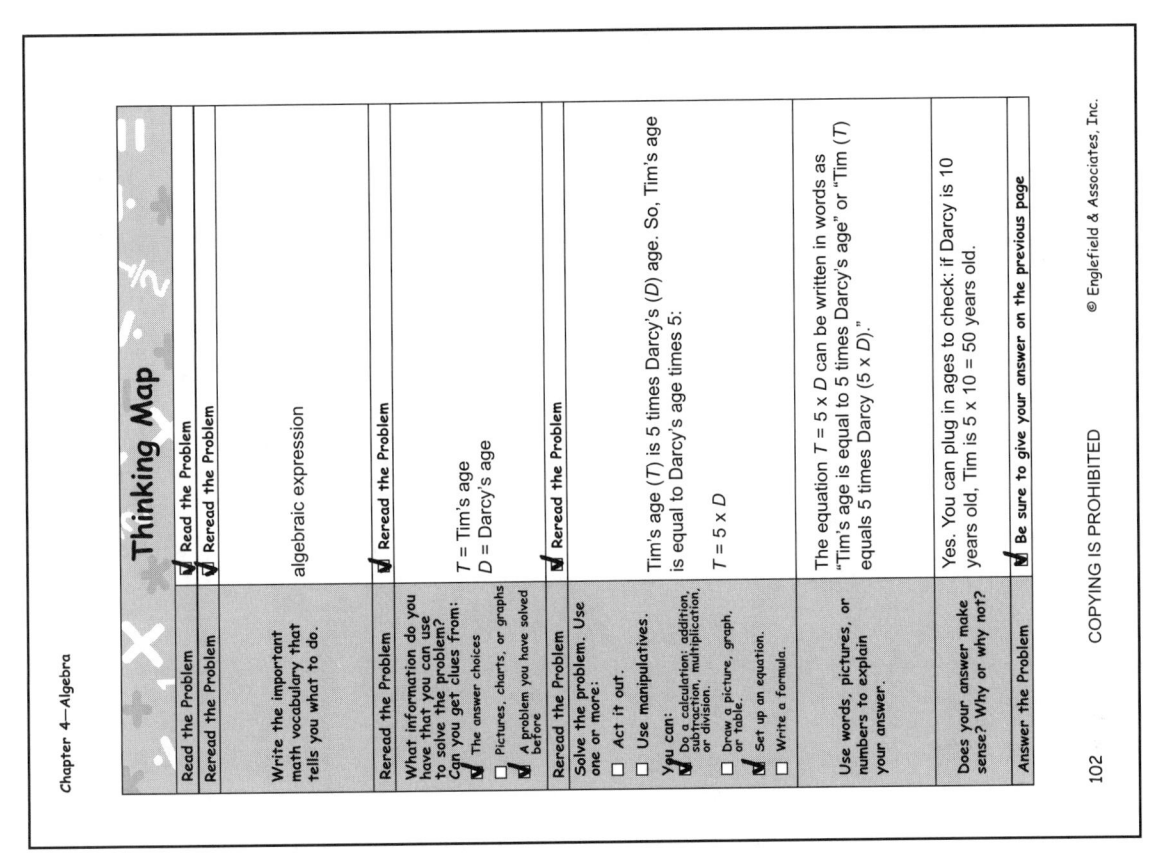

In the Student Workbook, the Thinking Map is blank.
In the Parent/Teacher Edition, suggested ways for completing the Thinking Maps are presented.

Algebra
Multiple-Choice Practice Problem 34

34. Tim is 5 times as old as Darcy. What is the correct algebraic expression to show Tim's age if T = Tim's age and D = Darcy's age?

○ A. $T \times 5 = D$
● B. $T = D \times 5$
○ C. $T = D + 5$
○ D. $T - 5 = D$

Use the thinking map on the next page to solve the problem.
Fill in the circle next to the correct answer.
Mark only one answer.

Chapter 4—Algebra

Thinking Map

☑ Read the Problem	
☑ Reread the Problem	half sheet full sheet >, <, = Which statement is true?
☑ Reread the Problem	
Write the important math vocabulary that tells you what to do.	
☑ Reread the Problem	
What information do you have that you can use to solve the problem? Can you get clues from: ☐ The answer choices ☑ Pictures, charts, or graphs ☐ A problem you have solved before	H = half sheet F = full sheet
☑ Reread the Problem	Draw a picture:
Solve the problem. Use one or more: ☐ Act it out. ☐ Use manipulatives. You can: ☐ Do a calculation: addition, subtraction, multiplication, or division. ☑ Draw a picture, graph, or table. ☐ Set up an equation. ☐ Write a formula.	$1/2$ of $1/2 = 1/4$ If you divide $1/2$ into two pieces ($1/2 \div 2$, or $1/2 \times 1/2$), each piece = $1/4$. $1/2 \times 1/2$ cake = $1/4$ cake $1/4 \times 1$ cake = $1/4$ cake
Use words, pictures, or numbers to explain your answer.	
Does your answer make sense? Why or why not?	Yes. $1/4$ cake + $1/4$ cake = $1/2$ cake. $1/4$ cake + $1/4$ cake + $1/4$ cake + $1/4$ cake = 1 cake.
☑ Answer the Problem	Be sure to give your answer on the previous page

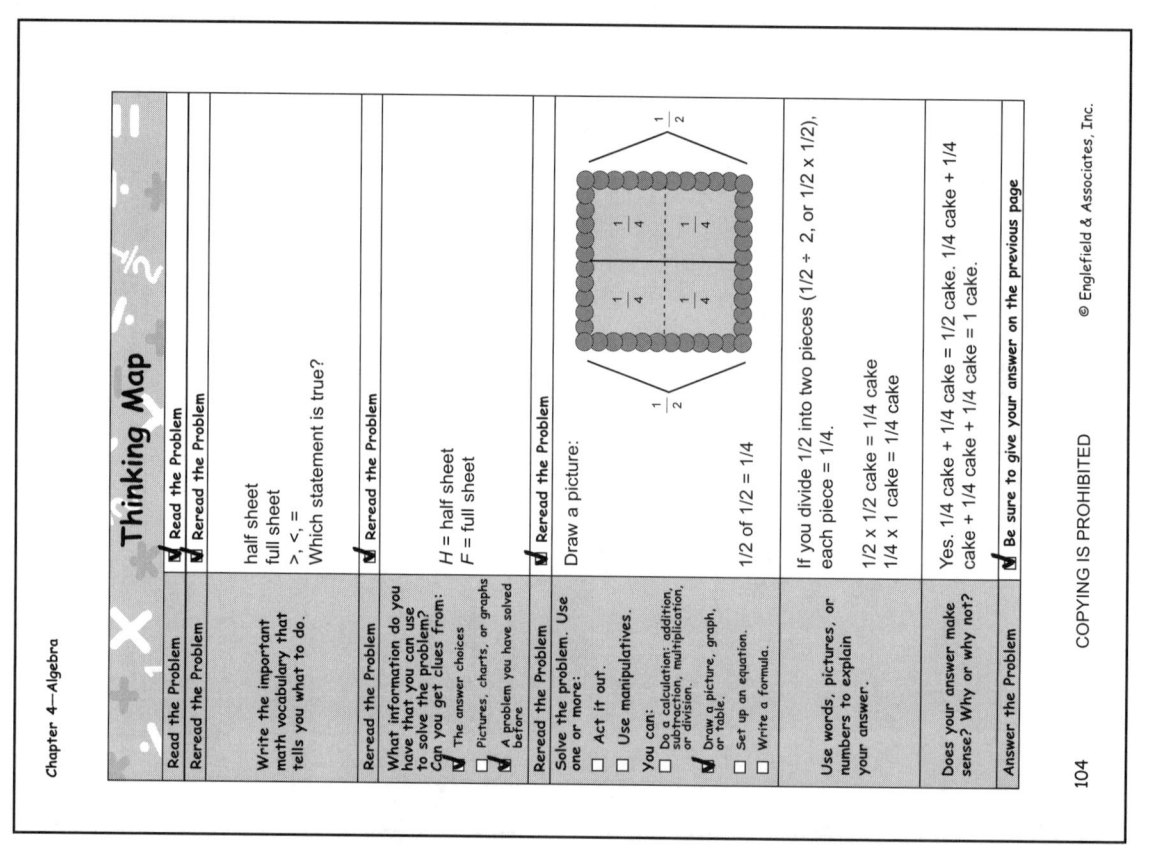

In the Student Workbook, the Thinking Map is blank.
In the Parent/Teacher Edition, suggested ways for completing the Thinking Maps are presented.

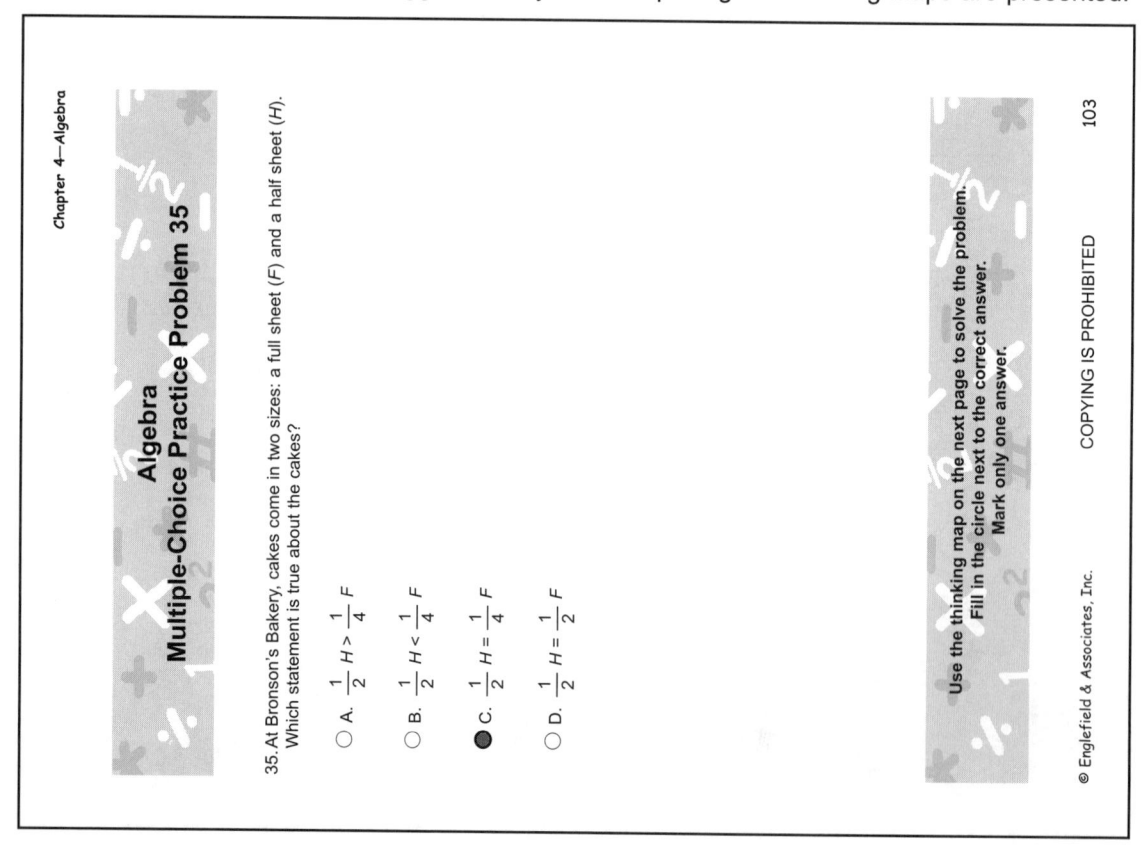

Algebra
Multiple-Choice Practice Problem 35

35. At Bronson's Bakery, cakes come in two sizes: a full sheet (F) and a half sheet (H). Which statement is true about the cakes?

○ A. $\frac{1}{2} H > \frac{1}{4} F$

○ B. $\frac{1}{2} H < \frac{1}{4} F$

● C. $\frac{1}{2} H = \frac{1}{4} F$

○ D. $\frac{1}{2} H = \frac{1}{2} F$

Use the thinking map on the next page to solve the problem.
Fill in the circle next to the correct answer.
Mark only one answer.

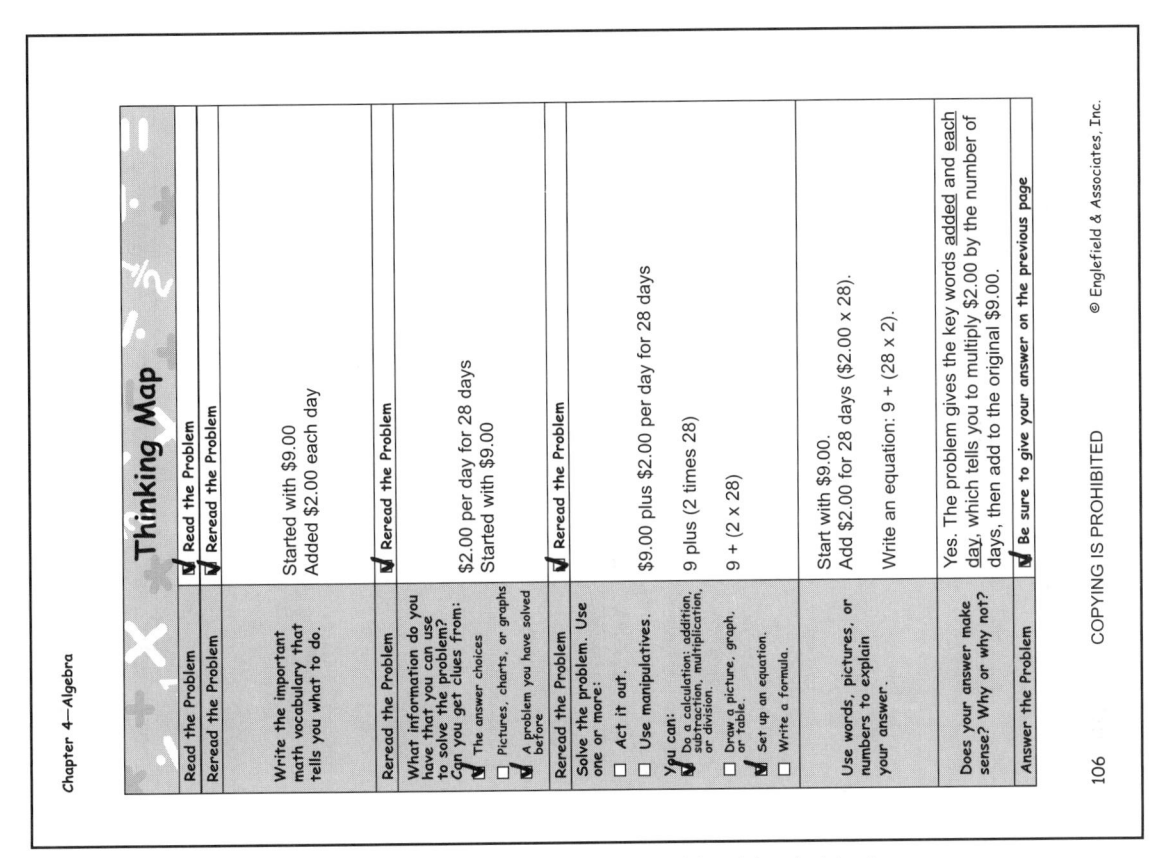

Thinking Map

Read the Problem	✓ Read the Problem
Reread the Problem	✓ Reread the Problem
Write the important math vocabulary that tells you what to do.	Started with $9.00 Added $2.00 each day
Reread the Problem	✓ Reread the Problem
What information do you have that you can use to solve the problem? Can you get clues from: ☐ The answer choices ☐ Pictures, charts, or graphs ✓ A problem you have solved before	$2.00 per day for 28 days Started with $9.00
Reread the Problem	✓ Reread the Problem
Solve the problem. Use one or more: ☐ Act it out. ☐ Use manipulatives. You can: ☐ Do a calculation: addition, subtraction, multiplication, or division. ☐ Draw a picture, graph, or table. ✓ Set up an equation. ☐ Write a formula.	$9.00 plus $2.00 per day for 28 days 9 plus (2 times 28) 9 + (2 × 28)
Use words, pictures, or numbers to explain your answer.	Start with $9.00. Add $2.00 for 28 days ($2.00 × 28). Write an equation: 9 + (28 × 2).
Does your answer make sense? Why or why not?	Yes. The problem gives the key words <u>added</u> and <u>each day</u>, which tells you to multiply $2.00 by the number of days, then add to the original $9.00.
Answer the Problem	✓ Be sure to give your answer on the previous page

---

Algebra
Multiple-Choice Practice Problem 36

36. Marin started with $9.00 in her bank account. Beginning on the first, she added $2.00 to her account each day for a month. How much money did she have on the 28th day of the month?

○ A. 28 + (9 × 2)
● B. 9 + (28 × 2)
○ C. 9 × (28 × 2)
○ D. (9 + 2) × 28

Use the thinking map on the next page to solve the problem.
Fill in the circle next to the correct answer.
Mark only one answer.

Chapter 4—Algebra

Thinking Map

Read the Problem ✓

Reread the Problem ✓

Write the important math vocabulary that tells you what to do.
— function

Reread the Problem ✓

What information do you have that you can use to solve the problem? Can you get clues from:
- ✓ The answer choices
- ☐ Pictures, charts, or graphs
- ☐ A problem you have solved before

In	Out
1/4	0.5
0.5	3/4
3/4	1
1	1 1/4

Reread the Problem ✓

Solve the problem. Use one or more:
- ☐ Act it out.
- ☐ Use manipulatives.

You can:
- ✓ Do a calculation: addition, subtraction, multiplication, or division.
- ✓ Draw a picture, graph, or table.
- ✓ Set up an equation.
- ☐ Write a formula.

Convert the decimals to fractions or the fractions to decimals and rewrite the function machine.

In	Out
1/4	2/4
2/4	3/4
3/4	4/4
4/4	1 1/4

1/4 + 1/4 = 2/4
2/4 + 1/4 = 3/4
3/4 + 1/4 = 4/4
4/4 + 1/4 = 1 1/4

In	Out
0.25	0.5
0.5	0.75
0.75	1
1	1.25

The function is to add 1/4. 0.25 = 1/4; therefore, the answer is A: Add 0.25.

Use words, pictures, or numbers to explain your answer.

Does your answer make sense? Why or why not?

Yes. The pattern is easy to see after you convert decimals to like fractions or fractions to decimals.

Answer the Problem ✓ Be sure to give your answer on the previous page.

108

In the Student Workbook, the Thinking Map is blank.
In the Parent/Teacher Edition, suggested ways for completing the Thinking Maps are presented.

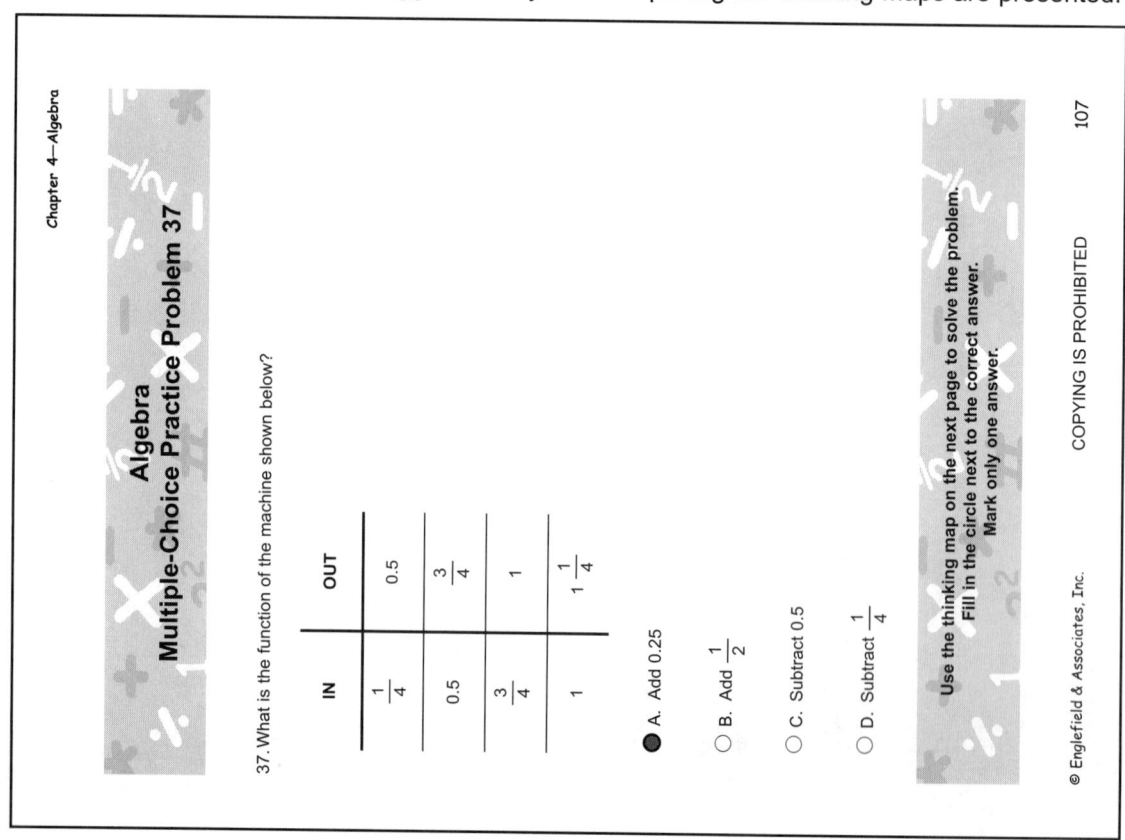

Algebra
Multiple-Choice Practice Problem 37

37. What is the function of the machine shown below?

IN	OUT
1/4	0.5
0.5	3/4
3/4	1
1	1 1/4

- ● A. Add 0.25
- ○ B. Add 1/2
- ○ C. Subtract 0.5
- ○ D. Subtract 1/4

Use the thinking map on the next page to solve the problem. Fill in the circle next to the correct answer. Mark only one answer.

Model Problem 11:
Algebra
Short-Answer

1. Prior Knowledge

Determine the student's prior knowledge about the strand, content of the problem, and/or mathematical process of the problem.

These questions can be used to help determine students' knowledge:

- What do you know about the commutative property?
- When do you see the commutative property used in real life?
- What tasks can you think of in which the completion does (or does not) require a certain order?

Additional prompts could be provided by using the following topics:

- putting on articles of clothing
- brushing your teeth
- washing and drying your clothes
- cooking from a recipe

Model Problem 11: Algebra Short-Answer

11. Use the commutative property of multiplication to write an equation equal to the equation below.

$$4Y \times \tfrac{1}{2} T \times 3Z = \square \times \square \times \square$$

> 1 point: The student demonstrates an understanding of the commutative property by changing the order of the numbers and variables.
>
> 1 point: The student correctly completes the equation.
>
> 1 point (extra credit): The student correctly lists all possible arrangements of the equation.

* For extra credit, list all of the possible answers.

Use the thinking map on the next page to solve the problem.
Fill in the circle next to the correct answer.
Mark only one answer.

Chapter 4—Algebra

2. Model

Model the use of the mathematical thinking map through class discussion with the thinking process modeled aloud as the thinking map is completed.

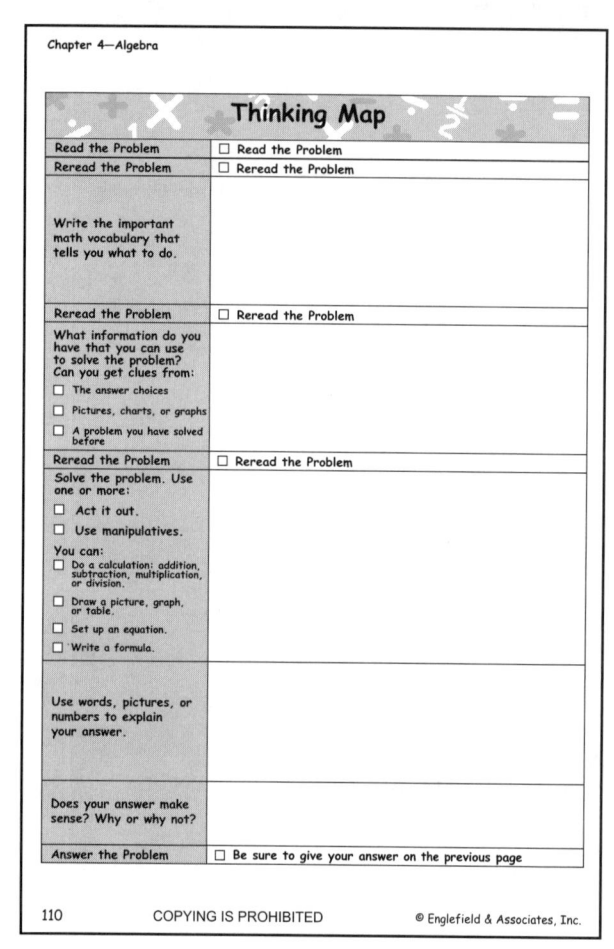

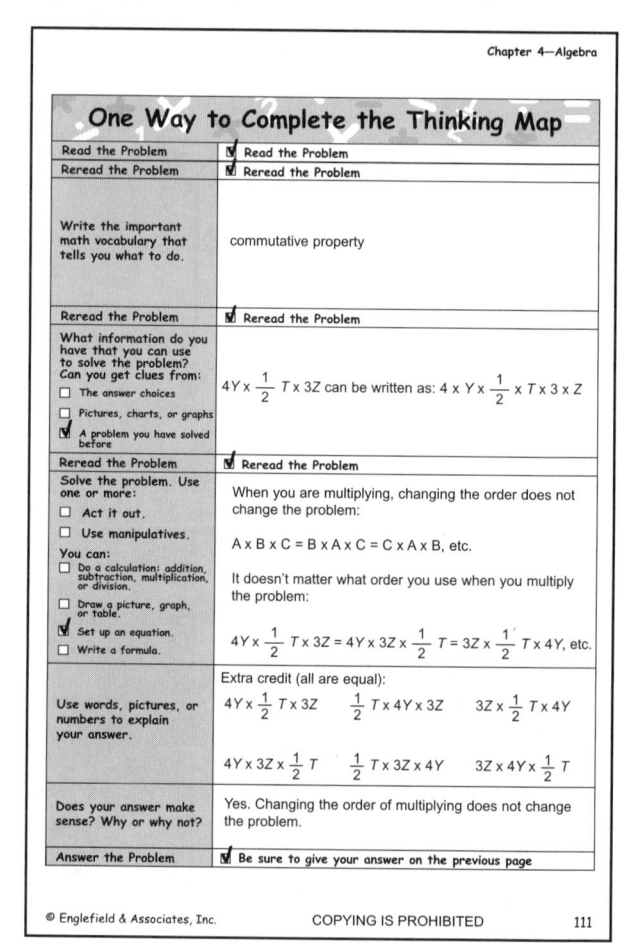

Thinking Map	
Read the Problem	☐ Read the Problem
Reread the Problem	☐ Reread the Problem
Write the important math vocabulary that tells you what to do.	
Reread the Problem	☐ Reread the Problem
What information do you have that you can use to solve the problem? Can you get clues from: ☐ The answer choices ☐ Pictures, charts, or graphs ☐ A problem you have solved before	
Reread the Problem	☐ Reread the Problem
Solve the problem. Use one or more: ☐ Act it out. ☐ Use manipulatives. You can: ☐ Do a calculation: addition, subtraction, multiplication, or division. ☐ Draw a picture, graph, or table. ☐ Set up an equation. ☐ Write a formula.	
Use words, pictures, or numbers to explain your answer.	
Does your answer make sense? Why or why not?	
Answer the Problem	☐ Be sure to give your answer on the previous page

One Way to Complete the Thinking Map	
Read the Problem	☑ Read the Problem
Reread the Problem	☑ Reread the Problem
Write the important math vocabulary that tells you what to do.	commutative property
Reread the Problem	☑ Reread the Problem
What information do you have that you can use to solve the problem? Can you get clues from: ☐ The answer choices ☐ Pictures, charts, or graphs ☑ A problem you have solved before	$4Y \times \frac{1}{2} T \times 3Z$ can be written as: $4 \times Y \times \frac{1}{2} \times T \times 3 \times Z$
Reread the Problem	☑ Reread the Problem
Solve the problem. Use one or more: ☐ Act it out. ☐ Use manipulatives. You can: ☐ Do a calculation: addition, subtraction, multiplication, or division. ☐ Draw a picture, graph, or table. ☑ Set up an equation. ☐ Write a formula.	When you are multiplying, changing the order does not change the problem: $A \times B \times C = B \times A \times C = C \times A \times B$, etc. It doesn't matter what order you use when you multiply the problem: $4Y \times \frac{1}{2} T \times 3Z = 4Y \times 3Z \times \frac{1}{2} T = 3Z \times \frac{1}{2} T \times 4Y$, etc.
Use words, pictures, or numbers to explain your answer.	Extra credit (all are equal): $4Y \times \frac{1}{2} T \times 3Z \quad \frac{1}{2} T \times 4Y \times 3Z \quad 3Z \times \frac{1}{2} T \times 4Y$ $4Y \times 3Z \times \frac{1}{2} T \quad \frac{1}{2} T \times 3Z \times 4Y \quad 3Z \times 4Y \times \frac{1}{2} T$
Does your answer make sense? Why or why not?	Yes. Changing the order of multiplying does not change the problem.
Answer the Problem	☑ Be sure to give your answer on the previous page

3. Guided Practice

Provide guided practice by presenting a problem for students to complete that matches the strand, content, or process you want to teach.

4. Independent Practice

Provide independent practice so students are provided with multiple-choice, short-answer, or extended-response problems.

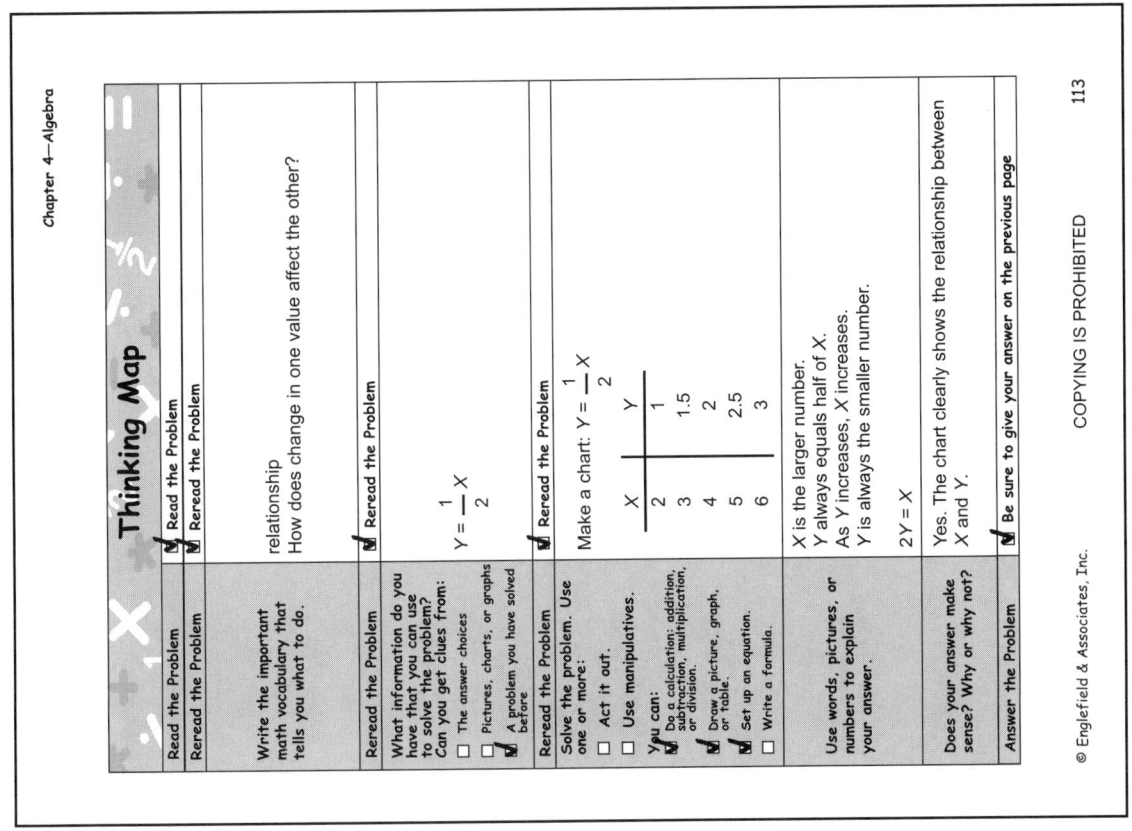

Chapter 4—Algebra

Thinking Map

Read the Problem	☑ Read the Problem
Reread the Problem	☑ Reread the Problem
Write the important math vocabulary that tells you what to do.	relationship How does change in one value affect the other?
	☑ Reread the Problem
What information do you have that you can use to solve the problem? Can you get clues from: ☐ The answer choices ☑ Pictures, charts, or graphs ☐ A problem you have solved before	$Y = \dfrac{1}{2} X$
	☑ Reread the Problem
Solve the problem. Use one or more: ☐ Act it out. ☐ Use manipulatives. You can: ☑ Do a calculation: addition, subtraction, multiplication, or division. ☑ Draw a picture, graph, or table. ☑ Set up an equation. ☐ Write a formula.	Make a chart: $Y = \dfrac{1}{2} X$ | X | Y | |---|---| | 2 | 1 | | 3 | 1.5 | | 4 | 2 | | 5 | 2.5 | | 6 | 3 | X is the larger number. Y always equals half of X. As Y increases, X increases. Y is always the smaller number.
Use words, pictures, or numbers to explain your answer.	$2Y = X$
Does your answer make sense? Why or why not?	Yes. The chart clearly shows the relationship between X and Y.
Answer the Problem	☑ Be sure to give your answer on the previous page

In the Student Workbook, the Thinking Map is blank.
In the Parent/Teacher Edition, suggested ways for completing the Thinking Maps are presented.

Algebra
Short-Answer Practice Problem 38

38. Use the equation below to answer the following questions.

$Y = \dfrac{1}{2} X$

What is the relationship between X and Y?

How does the change in one value affect the other?

---

1 point: The student determines that X is the larger number and has twice the value of Y.

1 point: The student demonstrates understanding that X will increase as Y increases, but Y will always equal one-half of X.

Use the thinking map on the next page to solve the problem.
Fill in the circle next to the correct answer.
Mark only one answer.

Chapter 4—Algebra

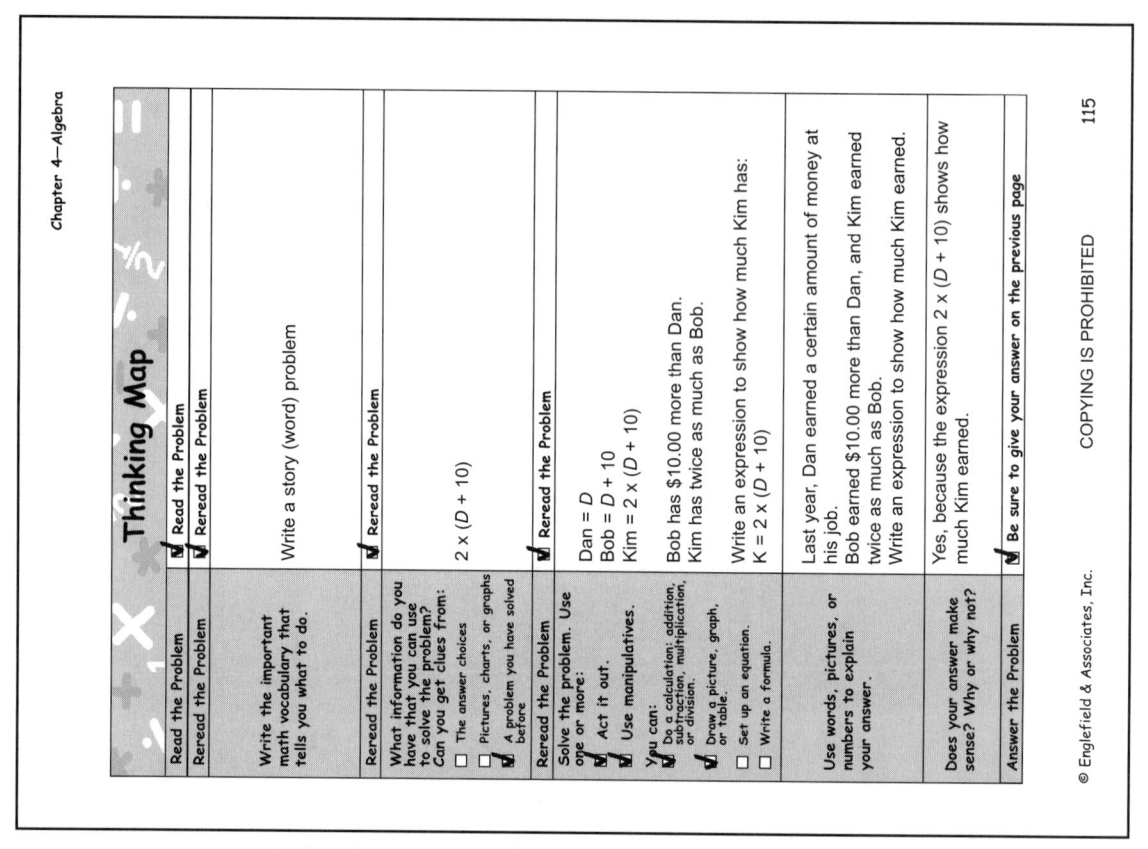

Thinking Map	
Read the Problem ☑ Read the Problem ☑ Reread the Problem	
Write the important math vocabulary that tells you what to do.	Write a story (word) problem
Reread the Problem ☑ Reread the Problem	
What information do you have that you can use to solve the problem? Can you get clues from: ☐ The answer choices ☐ Pictures, charts, or graphs ☐ A problem you have solved before	$2 \times (D + 10)$
Reread the Problem ☑ Reread the Problem	
Solve the problem. Use one or more: ☑ Act it out. ☑ Use manipulatives. You can: ☐ Do a calculation: addition, subtraction, multiplication, or division. ☑ Draw a picture, graph, or table. ☐ Set up an equation. ☐ Write a formula.	Dan = D Bob = D + 10 Kim = $2 \times (D + 10)$ Bob has $10.00 more than Dan. Kim has twice as much as Bob. Write an expression to show how much Kim has: K = $2 \times (D + 10)$
Use words, pictures, or numbers to explain your answer.	Last year, Dan earned a certain amount of money at his job. Bob earned $10.00 more than Dan, and Kim earned twice as much as Bob. Write an expression to show how much Kim earned.
Does your answer make sense? Why or why not?	Yes, because the expression $2 \times (D + 10)$ shows how much Kim earned.
Answer the Problem ☑ Be sure to give your answer on the previous page	

In the Student Workbook, the Thinking Map is blank.
In the Parent/Teacher Edition, suggested ways for completing the Thinking Maps are presented.

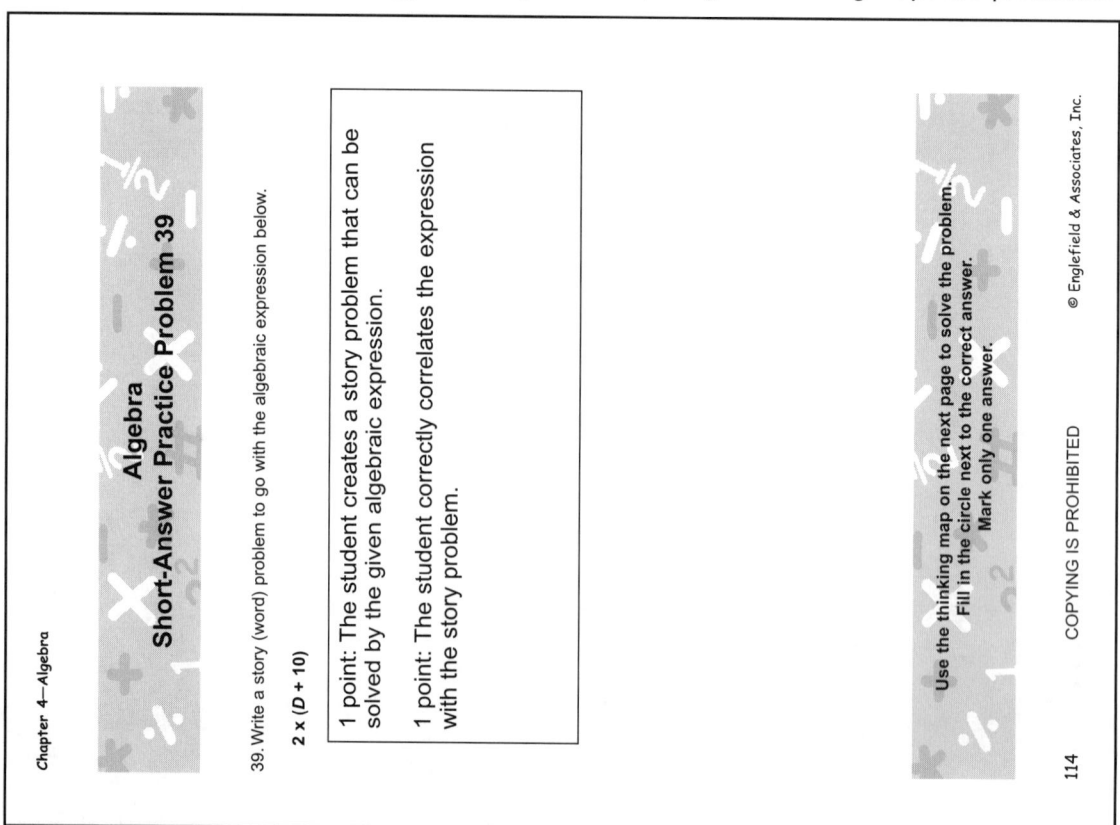

Algebra
Short-Answer Practice Problem 39

39. Write a story (word) problem to go with the algebraic expression below.

$2 \times (D + 10)$

1 point: The student creates a story problem that can be solved by the given algebraic expression.

1 point: The student correctly correlates the expression with the story problem.

Use the thinking map on the next page to solve the problem.
Fill in the circle next to the correct answer.
Mark only one answer.

Model Problem 12: Algebra Extended-Response

1. Prior Knowledge

Determine the student's prior knowledge about the strand, content of the problem, and/or mathematical process of the problem.

This question can be used to help determine students' knowledge:

- What do you know about solving for unknowns?
- When do you solve for unknowns in real-life situations?

Additional prompts could be provided by using the following topic:

- recipes
- money
- trip planning (bus or plane schedules)

Model Problem 12: Algebra Extended-Response

12. John weighs 35 pounds more than Erin does. Janet weighs 10 pounds less than Erin does. Together, the three weigh 340 pounds. How much does Erin weigh? Write an equation to solve the problem.

1 point: The student assigns a variable, such as E, to Erin.

1 point: The student determines that John = E + 35 and Janet = E − 10.

1 point: The student writes the correct equation: E + (E − 10) + (E + 35) = 340

1 point: The student simplifies the equation and finds Erin's weight:
3E + 25 = 340
E = 105.

Use the thinking map on the next page to solve the problem.
Fill in the circle next to the correct answer.
Mark only one answer.

Chapter 4—Algebra

2. Model

Model the use of the mathematical thinking map through class discussion with the thinking process modeled aloud as the thinking map is completed.

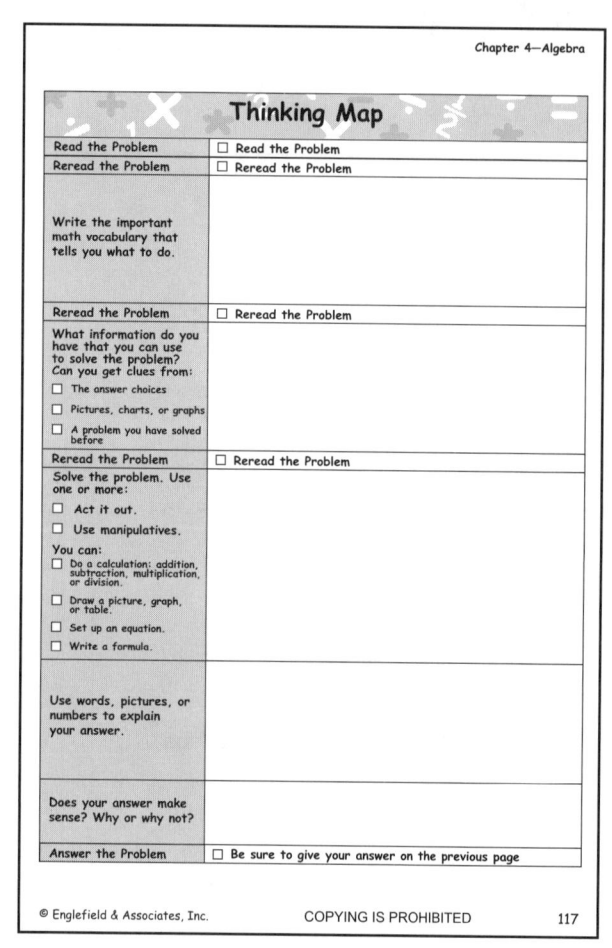

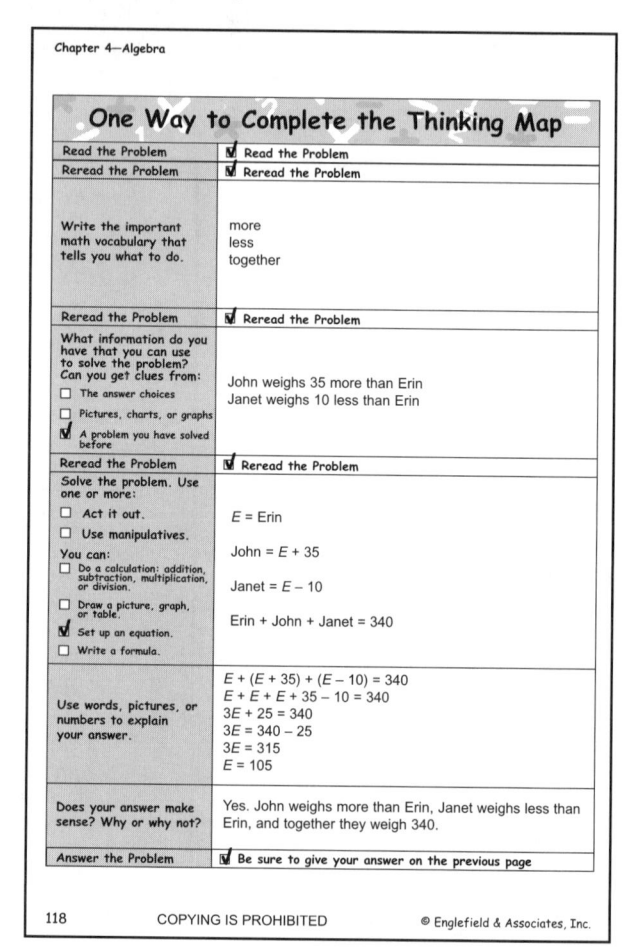

	Thinking Map
Read the Problem	☐ Read the Problem
Reread the Problem	☐ Reread the Problem
Write the important math vocabulary that tells you what to do.	
Reread the Problem	☐ Reread the Problem
What information do you have that you can use to solve the problem? Can you get clues from: ☐ The answer choices ☐ Pictures, charts, or graphs ☐ A problem you have solved before	
Reread the Problem	☐ Reread the Problem
Solve the problem. Use one or more: ☐ Act it out. ☐ Use manipulatives. You can: ☐ Do a calculation: addition, subtraction, multiplication, or division. ☐ Draw a picture, graph, or table. ☐ Set up an equation. ☐ Write a formula.	
Use words, pictures, or numbers to explain your answer.	
Does your answer make sense? Why or why not?	
Answer the Problem	☐ Be sure to give your answer on the previous page

	One Way to Complete the Thinking Map
Read the Problem	☑ Read the Problem
Reread the Problem	☑ Reread the Problem
Write the important math vocabulary that tells you what to do.	more less together
Reread the Problem	☑ Reread the Problem
What information do you have that you can use to solve the problem? Can you get clues from: ☐ The answer choices ☐ Pictures, charts, or graphs ☑ A problem you have solved before	John weighs 35 more than Erin Janet weighs 10 less than Erin
Reread the Problem	☑ Reread the Problem
Solve the problem. Use one or more: ☐ Act it out. ☐ Use manipulatives. You can: ☐ Do a calculation: addition, subtraction, multiplication, or division. ☐ Draw a picture, graph, or table. ☑ Set up an equation. ☐ Write a formula.	E = Erin John = $E + 35$ Janet = $E - 10$ Erin + John + Janet = 340
Use words, pictures, or numbers to explain your answer.	$E + (E + 35) + (E - 10) = 340$ $E + E + E + 35 - 10 = 340$ $3E + 25 = 340$ $3E = 340 - 25$ $3E = 315$ $E = 105$
Does your answer make sense? Why or why not?	Yes. John weighs more than Erin, Janet weighs less than Erin, and together they weigh 340.
Answer the Problem	☑ Be sure to give your answer on the previous page

3. Guided Practice

Provide guided practice by presenting a problem for students to complete that matches the strand, content, or process you want to teach.

4. Independent Practice

Provide independent practice so students are provided with multiple-choice, short-answer, or extended-response problems.

Thinking Map

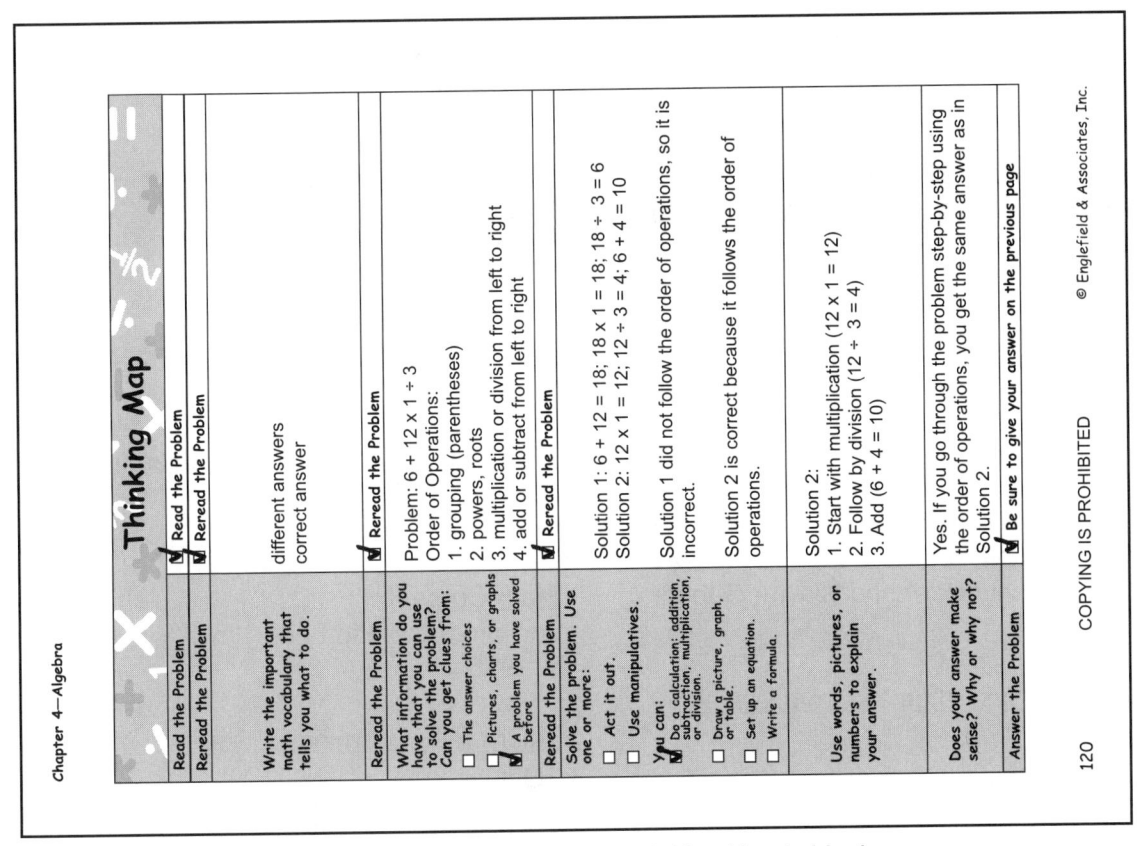

Read the Problem	☑ Read the Problem
Reread the Problem	☑ Reread the Problem
Write the important math vocabulary that tells you what to do.	different answers correct answer
Reread the Problem	☑ Reread the Problem
What information do you have that you can use to solve the problem? Can you get clues from: ☐ The answer choices ☐ Pictures, charts, or graphs ☑ A problem you have solved before	Problem: 6 + 12 x 1 ÷ 3 Order of Operations: 1. grouping (parentheses) 2. powers, roots 3. multiplication or division from left to right 4. add or subtract from left to right
Reread the Problem	☑ Reread the Problem
Solve the problem. Use one or more: ☐ Act it out. ☐ Use manipulatives. You can: ☑ Do a calculation: addition, subtraction, multiplication, or division. ☐ Draw a picture, graph, or table. ☐ Set up an equation. ☐ Write a formula.	Solution 1: 6 + 12 = 18; 18 x 1 = 18; 18 ÷ 3 = 6 Solution 2: 12 x 1 = 12; 12 ÷ 3 = 4; 6 + 4 = 10 Solution 1 did not follow the order of operations, so it is incorrect. Solution 2 is correct because it follows the order of operations.
Use words, pictures, or numbers to explain your answer.	Solution 2: 1. Start with multiplication (12 x 1 = 12) 2. Follow by division (12 ÷ 3 = 4) 3. Add (6 + 4 = 10)
Does your answer make sense? Why or why not?	Yes. If you go through the problem step-by-step using the order of operations, you get the same answer as in Solution 2.
Answer the Problem	☑ Be sure to give your answer on the previous page

In the Student Workbook, the Thinking Map is blank.
In the Parent/Teacher Edition, suggested ways for completing the Thinking Maps are presented.

Algebra
Extended-Response Practice Problem 40

40. Two students solved the same problem in different ways and ended with different answers. Show how each student may have solved the problem. Tell which way is correct and why the other way is incorrect. Explain your answer.

6 + 12 x 1 ÷ 3

1 point: The student determines that one person may have added 6 and 12, multiplied by 1, and then divided by 3, reaching an answer of 6.

1 point: The student determines that one person may have multiplied 12 by 1, divided by 3, and then added 6, reaching an answer of 10.

1 point: The student identifies 10 as the correct answer.

1 point: The student explains that the person who found 6 as a result did not follow the order of operations and the person who found 10 as a result did.

Use the thinking map on the next page to solve the problem.
Fill in the circle next to the correct answer.
Mark only one answer.

Chapter 4—Algebra

Literature Connections for Algebra

Pasta Math Problem Solving for "Alice in Pastaland": 40 Activities to Connect Math and Literature
Mary Chandler
© 1997

The Brain Explorer (Exploratorium at Home)
Pat Murphy
© 1999

Sir Cumference and the First Round Table: A Math Adventure
Cindy Neuschwander and Wayne Geehan
© 2002

More Sideways Arithmetic from Wayside School
Louis Sachar
© 1995

Among the Odds & Evens: A Tale of Adventure
Priscilla Turner
© 1999

Math Talk: Mathematics Ideas in Poems for Two Voices
Theoni Pappas
© 1991

If You Hopped Like a Frog
David M. Schwartz
© 1999

The King's Chessboard
David Birch
© 1993

Creative Puzzles of the World
Pieter Van Delft and Jack Botermans
© 1993

Alice in Pastaland: A Math Adventure
Alexandra Wright
© 1997

Math Curse
Jon Scieszka and Lane Smith
© 1995

Tiger Math: Learning to Graph from a Baby Tiger
Ann Whitehead Nagda
© 2000

The Rajah's Rice: A Mathematical Folktale from India
Adapted by David Barry (and others)
© 1994

Tiger Math: Learning to Graph from a Baby Tiger
Ann Whitehead Nagda and Cindy Bickel
© 2000

Chapter 5

Data and Probability

The purposes of this chapter include

1. Providing a definition of Data and Probability and what it means for fourth-grade students.
2. Offering teaching tips that reinforce Data and Probability concepts.
3. Offering multiple-choice, short-answer, and extended-response questions dealing with Data and Probability.
4. Providing ideas for connecting Data and Probability concepts and processes with other areas of the curriculum.

The following teaching tools are provided for extending students' thinking with Data and Probability:

- Model lesson for Data and Probability multiple-choice problems
- Seven Data and Probability multiple-choice practice problems
- Model lesson for Data and Probability short-answer problems
- Two Data and Probability short-answer practice problems
- Model lesson for Data and Probability extended-response problems
- One Data and Probability extended-response practice problem

What is Data and Probability?

In grade 4, students participate in projects or surveys that enable them to pose questions, collect data, represent data in a variety of ways, and to summarize findings to answer questions about their world. Students explore various means to portray data graphically. These include tables, line plots, picture graphs, bar graphs, or line graphs. They will also discover how data displays impact the way data is interpreted by the reader or audience. Students will formulate results based on their data evidence.

In probability, students will explore the likelihood of events to occur through experiments with spinners, cubes, or items chosen from a bag or a box.

Chapter 5—Data and Probability

What does Data and Probability look like?

- Students design investigations as they pose questions about their world, decide on data collection methods, collect data with observations, surveys, or experiments, and determine findings.
- Students will prepare data displays by selecting the best format to represent the data collected.
    - Picture, bar, or line graphs
    - Tables
    - Charts
    - Line plots
    - Timelines
- Students will experiment with the likelihood of events to occur.

Vocabulary related to Data and Probability:

Axis: A real or imaginary reference line. Graphs have two axes—the *x*-axis and the *y*-axis.

Bar graph: Uses bars to show quantities or numbers so they can be easily compared.

Box-and-whisker plot: A diagram used to show measures of central tendency, including the lowest value, the highest value, the lower quartile, the median, and the upper quartile in a set of data.

Chart: A sheet presenting information in the form of graphs or tables; a map on which specific information can be plotted.

Circle graph: A drawing or diagram, in the shape of a circle, that is used to record data in the form of percentages or ratios.

Conclusion: The result of an act or a process.

Continuous data: Data that can have an infinite number of possible values within a selected range.

Discrete data: Data that can only have a finite or limited number of possible values.

Double-bar graph: A type of bar graph used to compare two sets of similar data.

Fractional notation: A rational number of the form *a/b*; *a* is called the numerator (the number above the line in a fraction), and *b* is called the denominator (the number below the line in a fraction).

Frequency distribution: A table used in statistics as a method of recording the data collected. It lists a set of scores and their frequency, and a tally is often used to keep track of scores.

Graphs: Drawings or diagrams used to record information.

Interval: The distance between two points; the amount of time between two events.

Line graph: A linear drawing or diagram used to record information; it uses lines to join points representing the data.

Mean: The average of a number of different amounts, found by adding all of the amounts and dividing the total by the number of amounts that were added.

Median: The middle number or item in a set of numbers or objects arranged from least to greatest.

Mode: The number or object that appears most frequently in a set of numbers or objects. In a set of scores, the score that occurs the most.

Outcome: An end result.

Picture graph: A graph that uses pictures to represent quantities.

Probability: The chance that a particular outcome will occur, measured as a ratio of the number of favorable outcomes to the total number of possible outcomes.

Range: The difference between the greatest and least numbers in a set of data; from the lowest score to the highest score in a graph.

Ratio: The proportional value of two amounts found by dividing one by the other.

Stem-and-leaf plot: A diagram that compares data when numbers in a data set are close together. The "stem" is typically the digit in the tens place and the "leaves" are the digits in the ones place.

Survey: A questioning of a random sample of people to collect data or opinions.

Timeline: A schedule of activities or events; a representation of key events within a particular period; a chronology.

Venn diagram: A diagram that is used to show relationships between sets. Where the circles intersect, the sets contain the same elements.

Vertical bar graph or column graph: A graph that uses bars to show quantities or numbers so they can be easily compared.

Teaching Tips

- Students will examine issues in their classrooms, schools, or lives that they would like to investigate. Possible topics include:
  - How many children play on each piece of playground equipment during recess?
  - What lunch choices do students make?
  - What kinds and numbers of pets do fourth-grade students have?
- Students present their findings orally using visual representations prepared in class.
- Students formulate conclusions based on their findings.
- Students use spinners or cubes to predict the likelihood of a particular event occurring.

Chapter 5—Data and Probability

Model Problem 13: Data and Probability Multiple-Choice

1. Prior Knowledge

Determine the student's prior knowledge about the strand, content of the problem, and/or mathematical process of the problem.

These questions can be used to help determine students' knowledge:

- What do you know about probability?
- What do you know about rolling dice while playing a board game?

Additional prompts could be provided by using the following topics:

- flipping a coin
- spinning a spinner
- picking a name out of a hat
- being selected to participate in a survey or a research project
- medical or safety issues
- choosing clothing
- chances of a sports team winning

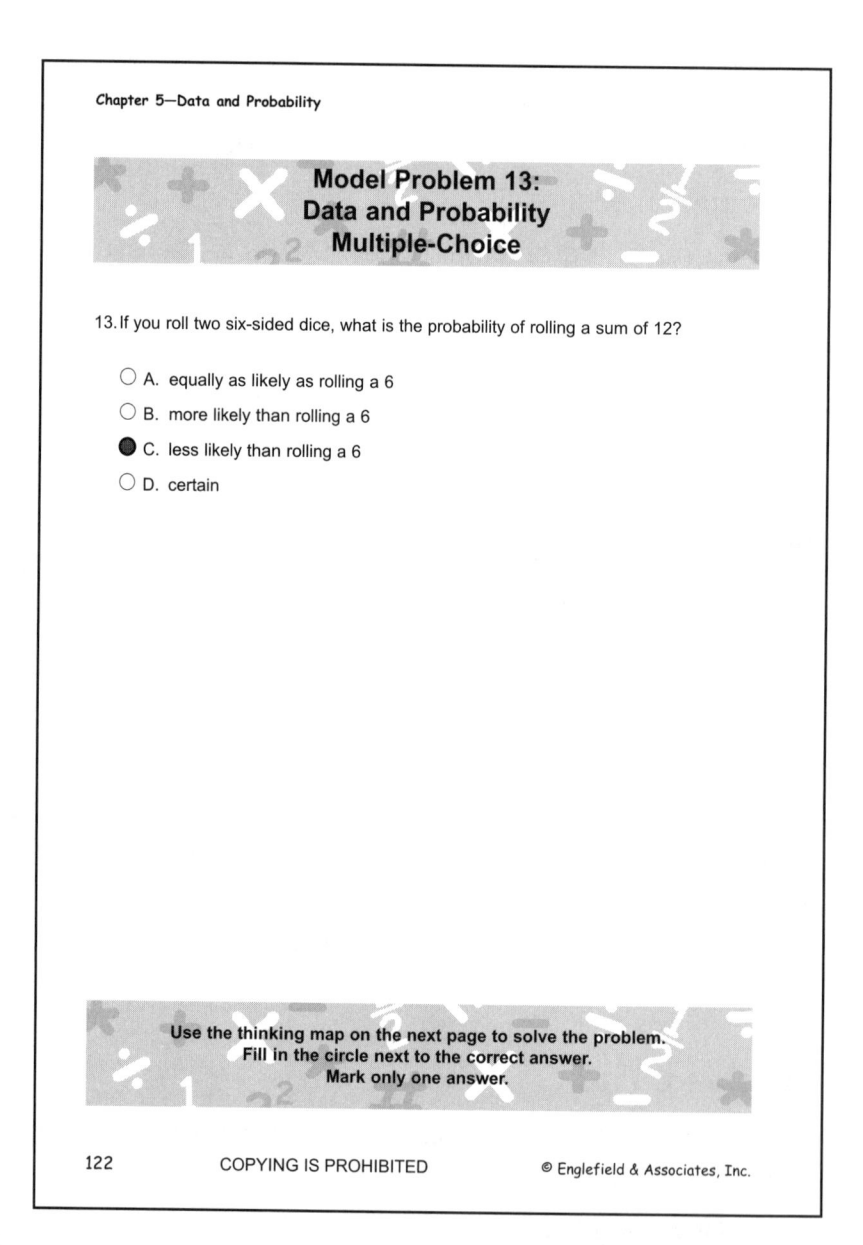

Chapter 5—Data and Probability

Model Problem 13: Data and Probability Multiple-Choice

13. If you roll two six-sided dice, what is the probability of rolling a sum of 12?

○ A. equally as likely as rolling a 6
○ B. more likely than rolling a 6
● C. less likely than rolling a 6
○ D. certain

Use the thinking map on the next page to solve the problem.
Fill in the circle next to the correct answer.
Mark only one answer.

2. Model

Model the use of the mathematical thinking map through class discussion with the thinking process modeled aloud as the thinking map is completed.

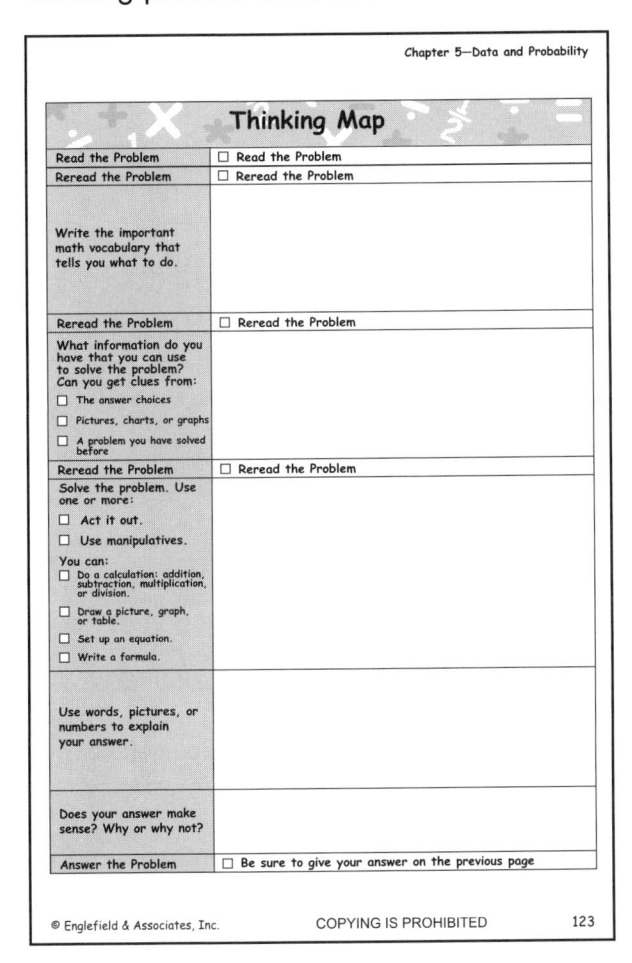

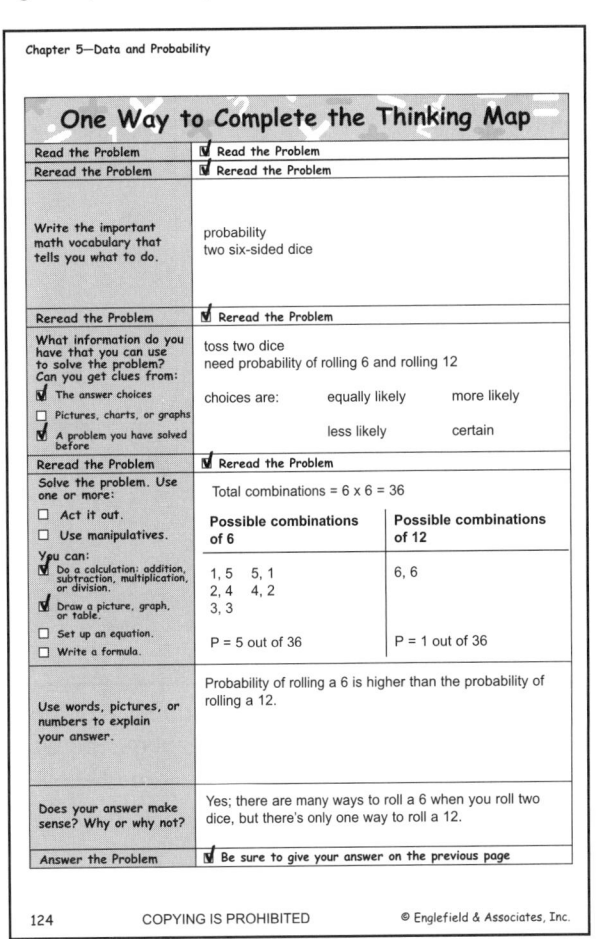

Thinking Map	One Way to Complete the Thinking Map
(blank template)	probability; two six-sided dice; toss two dice; need probability of rolling 6 and rolling 12; choices are: equally likely, more likely, less likely, certain; Total combinations = 6 x 6 = 36; Possible combinations of 6: 1,5 5,1 2,4 4,2 3,3; Possible combinations of 12: 6,6; P = 5 out of 36; P = 1 out of 36; Probability of rolling a 6 is higher than the probability of rolling a 12. Yes; there are many ways to roll a 6 when you roll two dice, but there's only one way to roll a 12.

3. Guided Practice

Provide guided practice by presenting a problem for students to complete that matches the strand, content, or process you want to teach.

4. Independent Practice

Provide independent practice so students are provided with multiple-choice, short-answer, or extended-response problems.

Chapter 5—Data and Probability

Thinking Map

Read the Problem ☑ Read the Problem ☑ Reread the Problem	
Write the important math vocabulary that tells you what to do.	missing score range
Reread the Problem ☑ Reread the Problem	
What information do you have that you can use to solve the problem? Can you get clues from: ☐ The answer choices ☑ Pictures, charts, or graphs ☐ A problem you have solved before	98 96 ___ 98 92 Range is 9
Reread the Problem ☑ Reread the Problem	
Solve the problem. Use one or more: ☐ Act it out. ☐ Use manipulatives. You can: ☑ Do a calculation: addition, subtraction, multiplication, or division. ☐ Draw a picture, graph, or table. ☑ Set up an equation. ☐ Write a formula.	By definition, the range is the difference between the high score and the low score. The highest and lowest scores given are 98 and 92: $98 - 92 = 6$. The range is 6. $9 - 6 = 3$ The missing score is 3 less than 92 or 3 more than 98.
Use words, pictures, or numbers to explain your answer.	$92 - 3 = 89$ $98 + 3 = 101$ The score must be less than 100, so 89 must be the missing score $98 - 89 = 9$
Does your answer make sense? Why or why not?	Yes, because the difference between the highest and lowest scores given is 6, so the missing score had to be less than 92.
Answer the Problem ☑ Be sure to give your answer on the previous page	

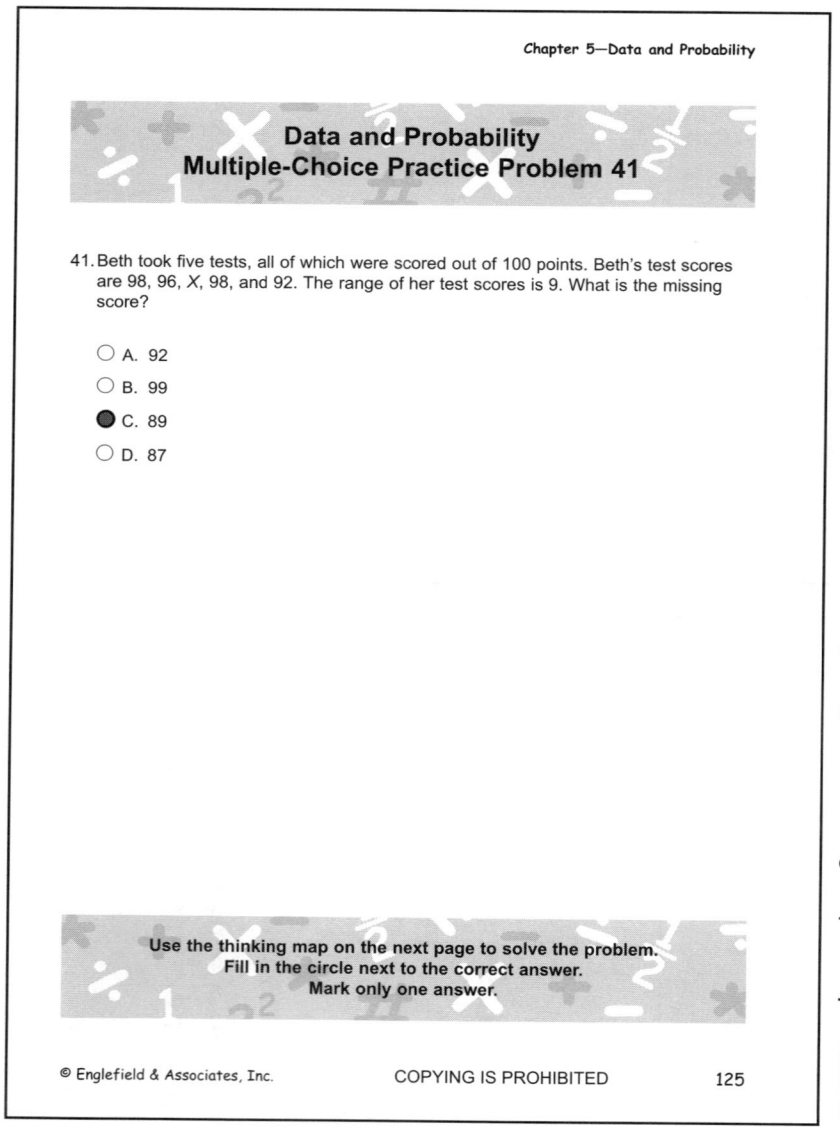

126

In the Student Workbook, the Thinking Map is blank.
In the Parent/Teacher Edition, suggested ways for completing the Thinking Maps are presented.

Data and Probability
Multiple-Choice Practice Problem 41

41. Beth took five tests, all of which were scored out of 100 points. Beth's test scores are 98, 96, X, 98, and 92. The range of her test scores is 9. What is the missing score?

○ A. 92
○ B. 99
● C. 89
○ D. 87

Use the thinking map on the next page to solve the problem.
Fill in the circle next to the correct answer.
Mark only one answer.

125

Chapter 5—Data and Probability

Thinking Map

✓ **Read the Problem**	
✓ **Reread the Problem**	
Write the important math vocabulary that tells you what to do.	mode How many were collected?
✓ **Reread the Problem**	
What information do you have that you can use to solve the problem? Can you get clues from: ☑ The answer choices ☐ Pictures, charts, or graphs ☑ A problem you have solved before	Mon Tues Wed Thurs Fri 16 32 ? 21 20
✓ **Reread the Problem**	
Solve the problem. Use one or more: ☐ Act it out. ☐ Use manipulatives. You can: ☐ Do a calculation: addition, subtraction, multiplication, or division. ☐ Draw a picture, graph, or table. ☐ Set up an equation. ☐ Write a formula.	Definition of mode: The value that occurs most often. Since each score occurs once, the missing score must be 21.
Use words, pictures, or numbers to explain your answer.	Value that occurs most: 16, 20, 21, 32, ? For the mode to be 21, the missing score has to be 21.
Does your answer make sense? Why or why not?	Yes—if I replace 21 in the blank, the mode of the set will be 21.
Answer the Problem	
✓ Be sure to give your answer on the previous page	

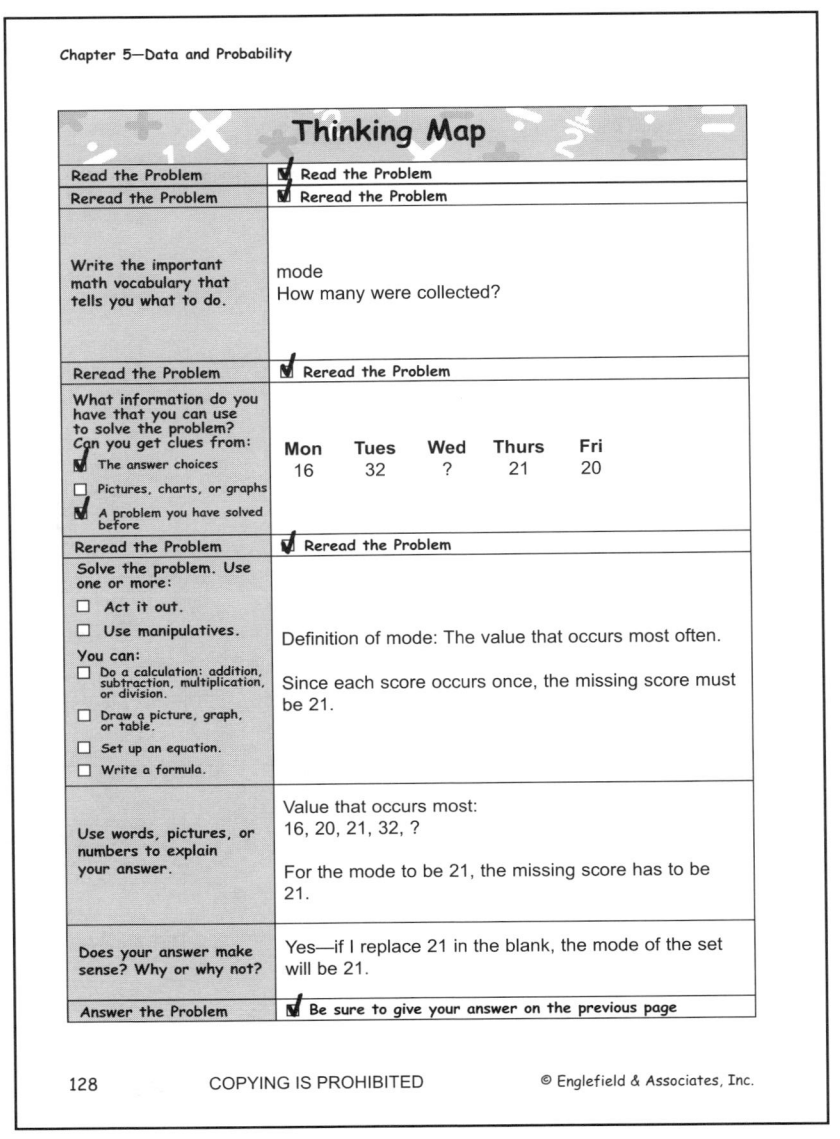

128

In the Student Workbook, the Thinking Map is blank.
In the Parent/Teacher Edition, suggested ways for completing the Thinking Maps are presented.

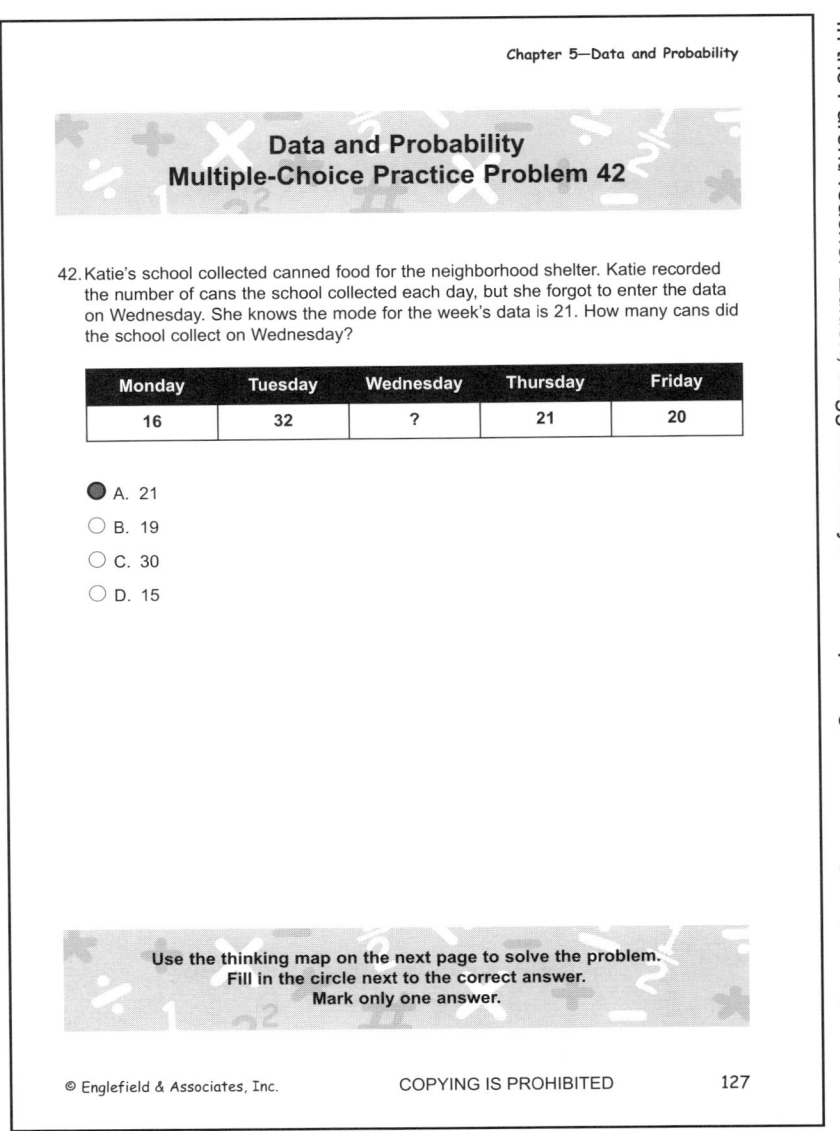

Data and Probability
Multiple-Choice Practice Problem 42

42. Katie's school collected canned food for the neighborhood shelter. Katie recorded the number of cans the school collected each day, but she forgot to enter the data on Wednesday. She knows the mode for the week's data is 21. How many cans did the school collect on Wednesday?

Monday	Tuesday	Wednesday	Thursday	Friday
16	32	?	21	20

● A. 21
○ B. 19
○ C. 30
○ D. 15

Use the thinking map on the next page to solve the problem.
Fill in the circle next to the correct answer.
Mark only one answer.

127

Chapter 5—Data and Probability

Thinking Map

Read the Problem	☑ Read the Problem
Reread the Problem	☑ Reread the Problem
Write the important math vocabulary that tells you what to do.	compare type of graph
Reread the Problem	☑ Reread the Problem
What information do you have that you can use to solve the problem? Can you get clues from: ☐ The answer choices ☐ Pictures, charts, or graphs ☐ A problem you have solved before	Bar graph Line graph Double-bar graph Stem-and-leaf plot
Reread the Problem	☑ Reread the Problem
Solve the problem. Use one or more: ☐ Act it out. ☐ Use manipulatives. You can: ☐ Do a calculation: addition, subtraction, multiplication, or division. ☐ Draw a picture, graph, or table. ☐ Set up an equation. ☐ Write a formula.	By definition: Bar graph – compares data. Line graph – shows ups and downs in trends. Double-bar graph – compares 2 sets of data. Stem-and-leaf plot – compares data when numbers are close together in value; doesn't label individual values.
Use words, pictures, or numbers to explain your answer.	[bar graph showing Bob, Sue, Joyce, Tyrone, Pete, Cindy]
Does your answer make sense? Why or why not?	Yes. The bar graph is a good way to display one set of data.
Answer the Problem	☑ Be sure to give your answer on the previous page

In the Student Workbook, the Thinking Map is blank.
In the Parent/Teacher Edition, suggested ways for completing the Thinking Maps are presented.

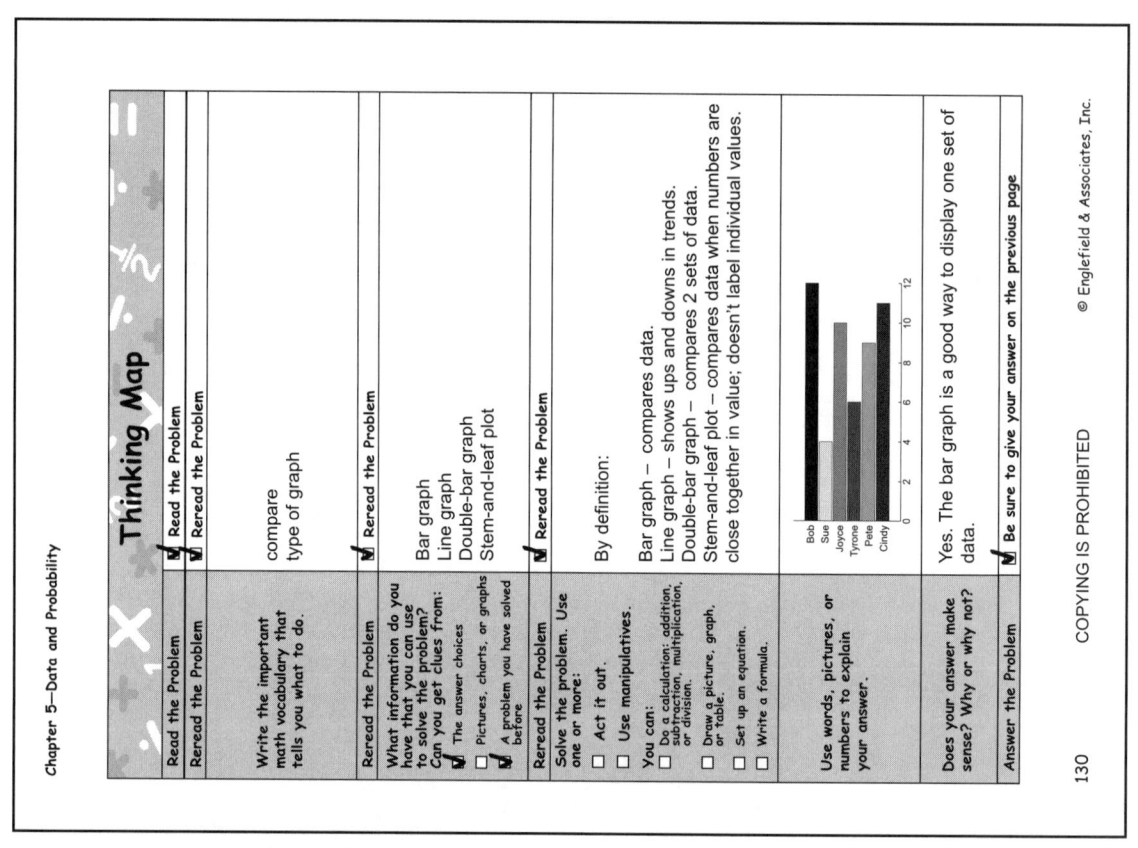

Data and Probability
Multiple-Choice Practice Problem 43

43. The table below shows how many chess games each player won in the school tournament.

NAME	WINS
Bob	12
Sue	4
Joyce	10
Tyrone	6
Pete	9
Cindy	11

If you wanted to compare the number of wins by each player, what type of graph would best display the data?

● A. bar graph
○ B. line graph
○ C. double-bar graph
○ D. stem-and-leaf plot

Use the thinking map on the next page to solve the problem
Fill in the circle next to the correct answer.
Mark only one answer.

Chapter 5—Data and Probability

Thinking Map

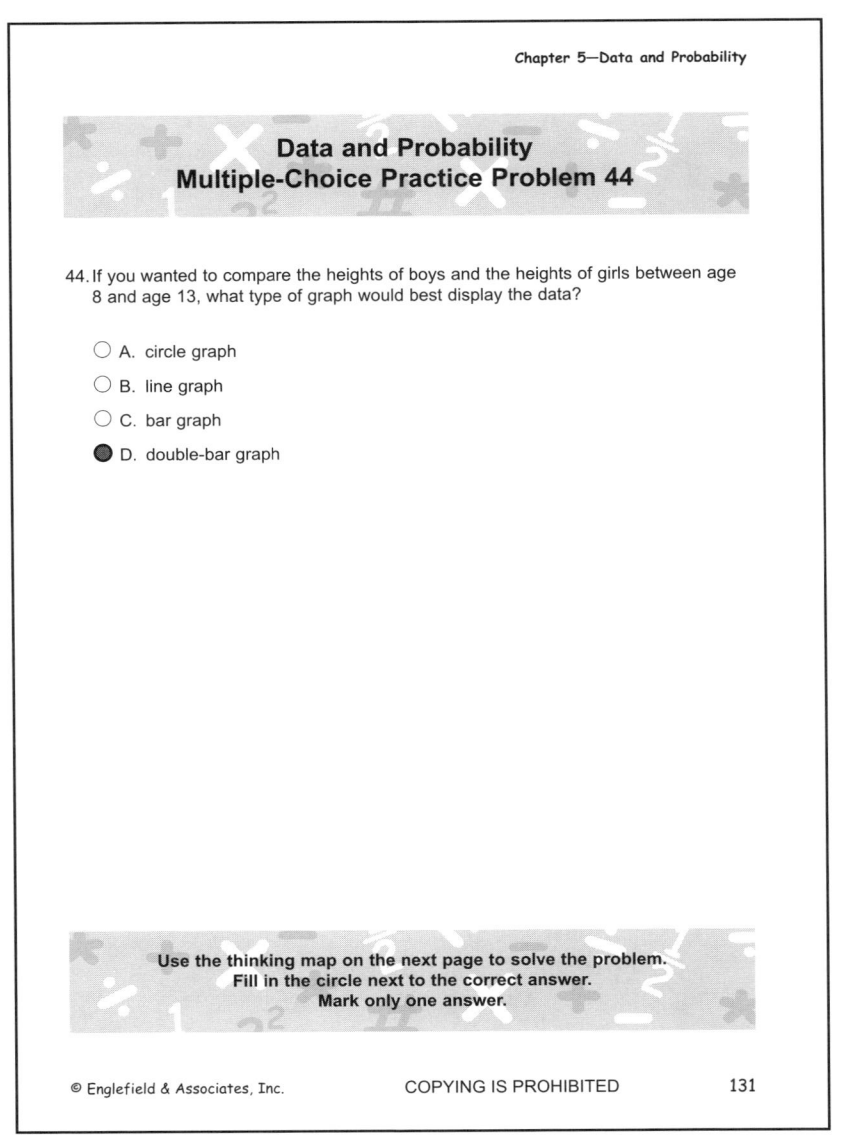

☑ **Read the Problem**	
☑ **Reread the Problem**	
Write the important math vocabulary that tells you what to do.	compare data between boys and girls height between 8 and 13
☑ **Reread the Problem**	
What information do you have that you can use to solve the problem? Can you get clues from: ☐ The answer choices ☑ Pictures, charts, or graphs ☑ A problem you have solved before	circle graph line graph bar graph double bar graph
☑ **Reread the Problem**	
Solve the problem. Use one or more: ☐ Act it out. ☐ Use manipulatives. **You can:** ☐ Do a calculation: addition, subtraction, multiplication, or division. ☑ Draw a picture, graph, or table. ☐ Set up an equation. ☐ Write a formula.	Define answer choices: Circle graph – shows percentages or parts of a whole. Line graph – shows ups and downs in trends. Bar graph – compares data. Double bar graph – good for comparing quantities between sets. [double bar graph showing girls and boys heights at ages 8, 9, 10, 11, 12, 13 with y-axis 0.5' to 5.5'] Since the comparison shows the heights at different ages for both boys and girls, a double bar graph would be the best choice.
Use words, pictures, or numbers to explain your answer.	
Does your answer make sense? Why or why not?	Yes; the double bar graph can compare the heights of both boys and girls at different ages.
Answer the Problem	☑ Be sure to give your answer on the previous page

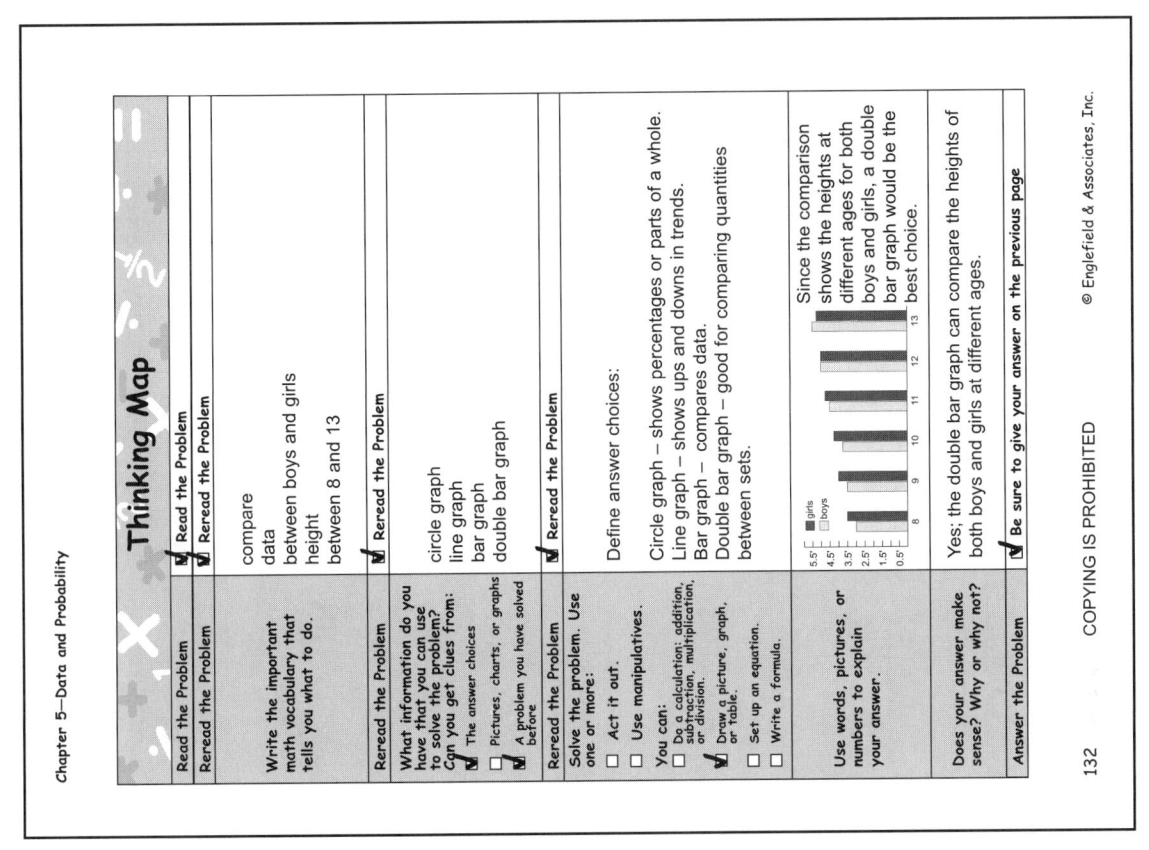

132

In the Student Workbook, the Thinking Map is blank.
In the Parent/Teacher Edition, suggested ways for completing the Thinking Maps are presented.

Data and Probability
Multiple-Choice Practice Problem 44

44. If you wanted to compare the heights of boys and the heights of girls between age 8 and age 13, what type of graph would best display the data?

○ A. circle graph

○ B. line graph

○ C. bar graph

● D. double-bar graph

Use the thinking map on the next page to solve the problem.
Fill in the circle next to the correct answer.
Mark only one answer.

131

Chapter 5—Data and Probability

Thinking Map

Read the Problem	✓ Read the Problem ✓ Reread the Problem
Write the important math vocabulary that tells you what to do.	probability spinner
Reread the Problem	✓ Reread the Problem
What information do you have that you can use to solve the problem? Can you get clues from: ☐ The answer choices ✓ Pictures, charts, or graphs ☐ A problem you have solved before	
Reread the Problem	✓ Reread the Problem
Solve the problem. Use one or more: ✓ Act it out. ✓ Use manipulatives. You can: ☐ Do a calculation: addition, subtraction, multiplication, or division. ☐ Draw a picture, graph, or table. ☐ Set up an equation. ☐ Write a formula.	Try experimenting with a spinner to help determine the probability. If you spin this spinner once, you have three chances to land on red and two chances to land on blue. There are six chances altogether. 3/6 chances for red + 2/6 chances for blue = 5/6 for either red or blue.
Use words, pictures, or numbers to explain your answer.	Add the probability for red and the probability for blue: 3/6 + 2/6 = 5/6 5/6 = Probability for red or blue
Does your answer make sense? Why or why not?	Yes. Five out of the six sections are either red or blue; only one out of six is another color (green).
Answer the Problem	✓ Be sure to give your answer on the previous page

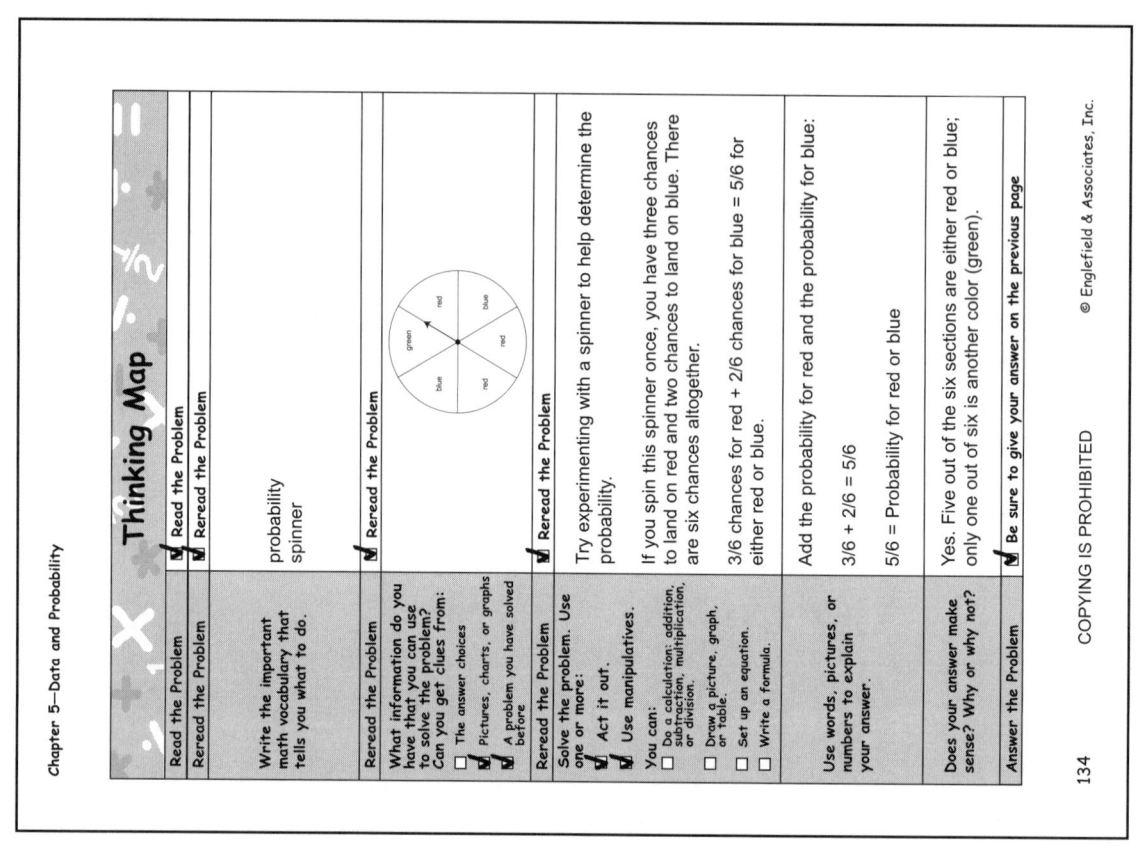

In the Student Workbook, the Thinking Map is blank.
In the Parent/Teacher Edition, suggested ways for completing the Thinking Maps are presented.

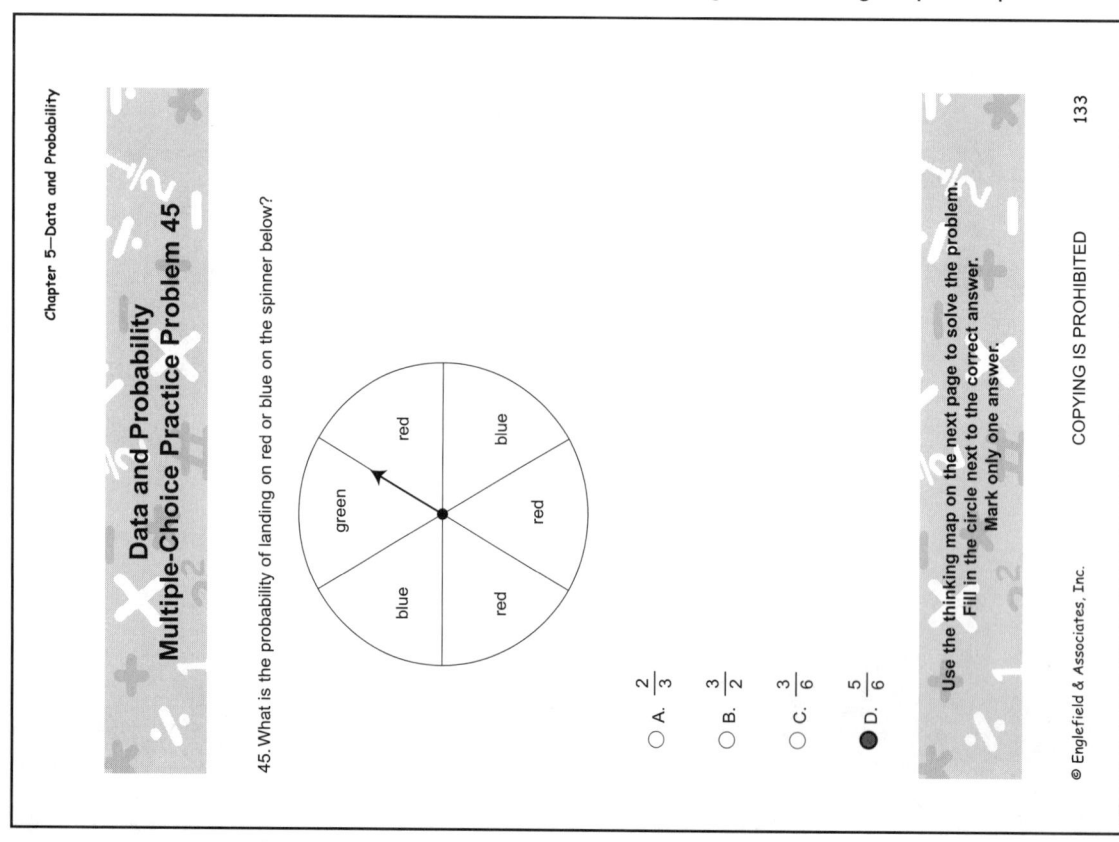

Data and Probability
Multiple-Choice Practice Problem 45

45. What is the probability of landing on red or blue on the spinner below?

○ A. 2/3
○ B. 3/2
○ C. 3/6
● D. 5/6

Use the thinking map on the next page to solve the problem.
Fill in the circle next to the correct answer.
Mark only one answer.

Thinking Map

Read the Problem ✓	
Reread the Problem ✓	How many combinations?
Write the important math vocabulary that tells you what to do.	
Reread the Problem ✓	3 sizes: small, medium, large
What information do you have that you can use to solve the problem? Can you get clues from:	3 items: sweatshirts, sweatpants, T-shirts
☐ The answer choices	2 color choices
☐ Pictures, charts, or graphs	
☐ A problem you have solved before	
Reread the Problem ✓	small pants: color 1, color 2
Solve the problem. Use one or more:	small T-shirt: color 1, color 2
☐ Act it out.	med. pants: color 1, color 2
☐ Use manipulatives.	med. T-shirt: color 1, color 2
You can:	large pants: color 1
☐ Do a calculation: addition, subtraction, multiplication, or division.	large T-shirt: color 1, color 2 Etc.
✓ Draw a picture, graph, or table.	6 pants choices, 6 sweatshirt choices, 6 T-shirt choices
☐ Set up an equation.	sweatpants: 3 sizes x 2 colors = 6 choices
☐ Write a formula.	sweatshirt: 3 sizes x 2 colors = 6 choices
	T-shirt: 3 sizes x 2 colors = 6 choices
Use words, pictures, or numbers to explain your answer.	6 choices + 6 choices + 6 choices = 18 choices
Does your answer make sense? Why or why not?	Yes, because if you write out all the choices for each item and count them, there are 18.
Answer the Problem ✓	Be sure to give your answer on the previous page

In the Student Workbook, the Thinking Map is blank.
In the Parent/Teacher Edition, suggested ways for completing the Thinking Maps are presented.

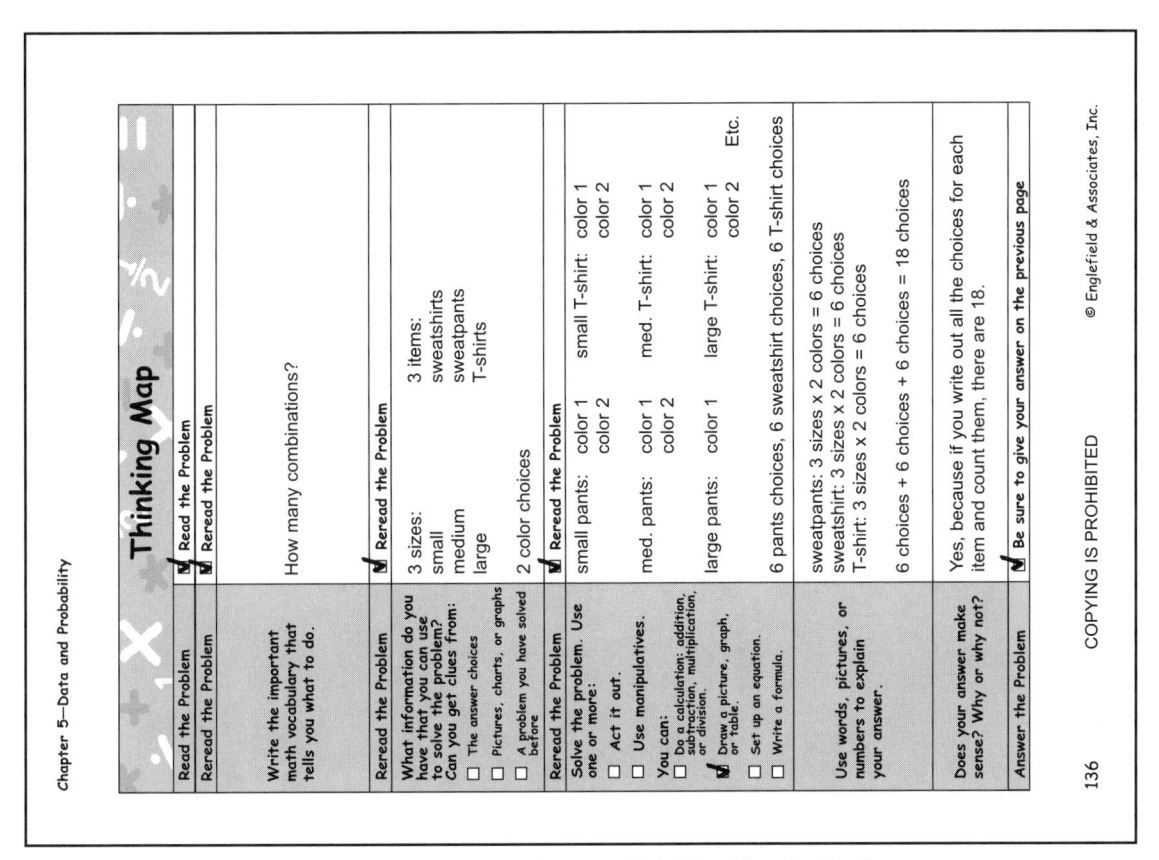

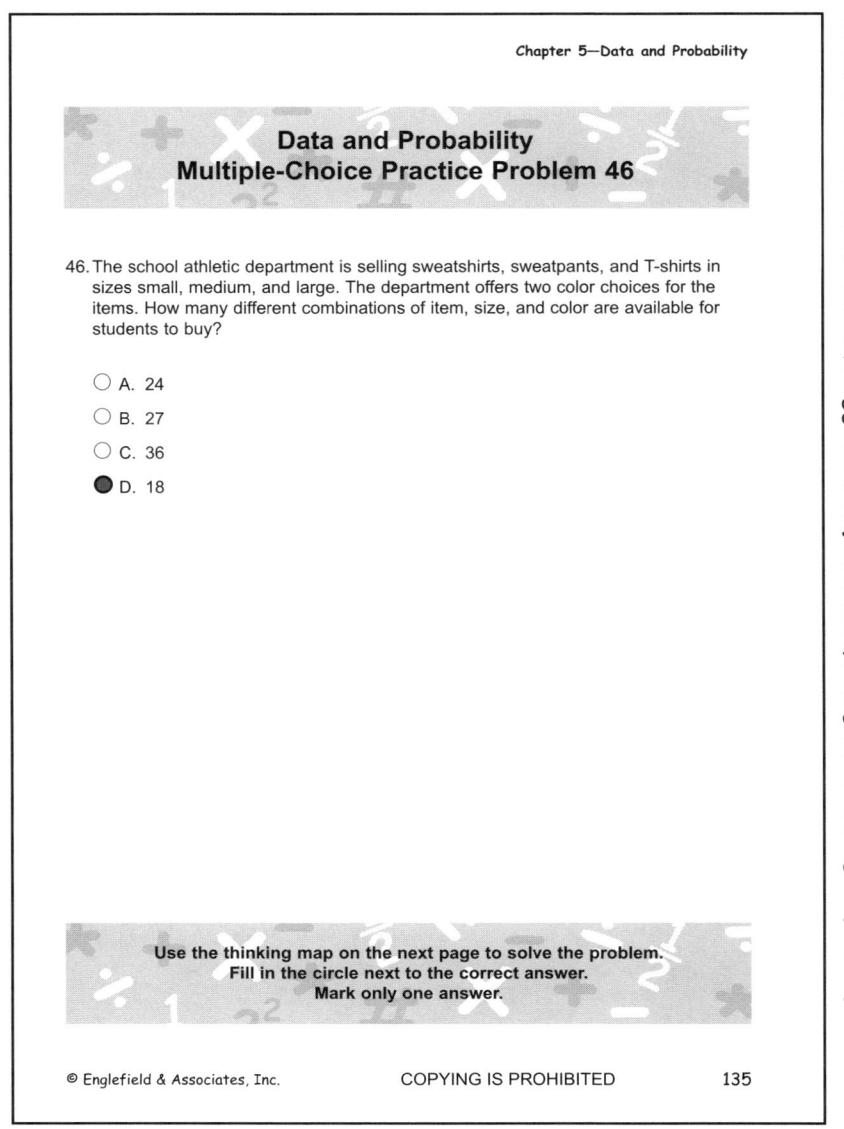

Data and Probability
Multiple-Choice Practice Problem 46

46. The school athletic department is selling sweatshirts, sweatpants, and T-shirts in sizes small, medium, and large. The department offers two color choices for the items. How many different combinations of item, size, and color are available for students to buy?

○ A. 24
○ B. 27
○ C. 36
● D. 18

Use the thinking map on the next page to solve the problem.
Fill in the circle next to the correct answer.
Mark only one answer.

Chapter 5—Data and Probability

Thinking Map

Read the Problem	☑ Read the Problem
	☑ Reread the Problem
Write the important math vocabulary that tells you what to do.	What are the chances? likely = greater than 50% unlikely = less than 50% equal = 50% certain = =100%
Reread the Problem	☑ Reread the Problem
What information do you have that you can use to solve the problem? Can you get clues from: ☐ The answer choices ☑ Pictures, charts, or graphs ☑ A problem you have solved before	2 pennies 2 dimes What are the chances of drawing one dime?
Reread the Problem	☑ Reread the Problem
Solve the problem. Use one or more: ☑ Act it out. ☐ Use manipulatives. You can: ☐ Do a calculation: addition, subtraction, multiplication, or division. ☑ Draw a picture, graph, or table. ☐ Set up an equation. ☐ Write a formula.	4 options: 1 dime, 1 dime, 1 penny, 1 penny Options for drawing 1 dime = 2 likely = greater than 50% unlikely = less than 50% equal = 50% certain = 100% 2 chances to draw a dime out of 4 total chances. 2/4 = 1/2 = 50% or equal chance
Use words, pictures, or numbers to explain your answer.	
Does your answer make sense? Why or why not?	Yes: 2/4 = 1/2 = 50% chance, which means it is equally likely.
Answer the Problem	☑ Be sure to give your answer on the previous page

In the Student Workbook, the Thinking Map is blank.
In the Parent/Teacher Edition, suggested ways for completing the Thinking Maps are presented.

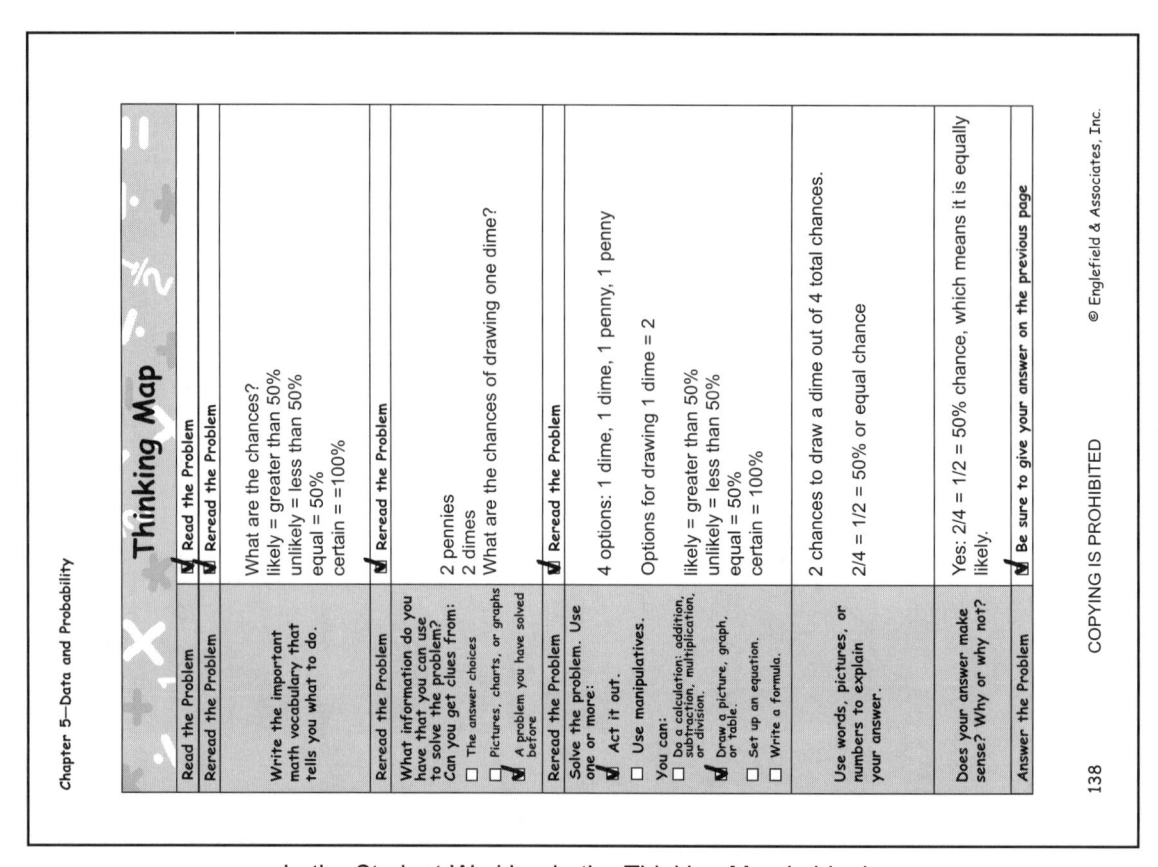

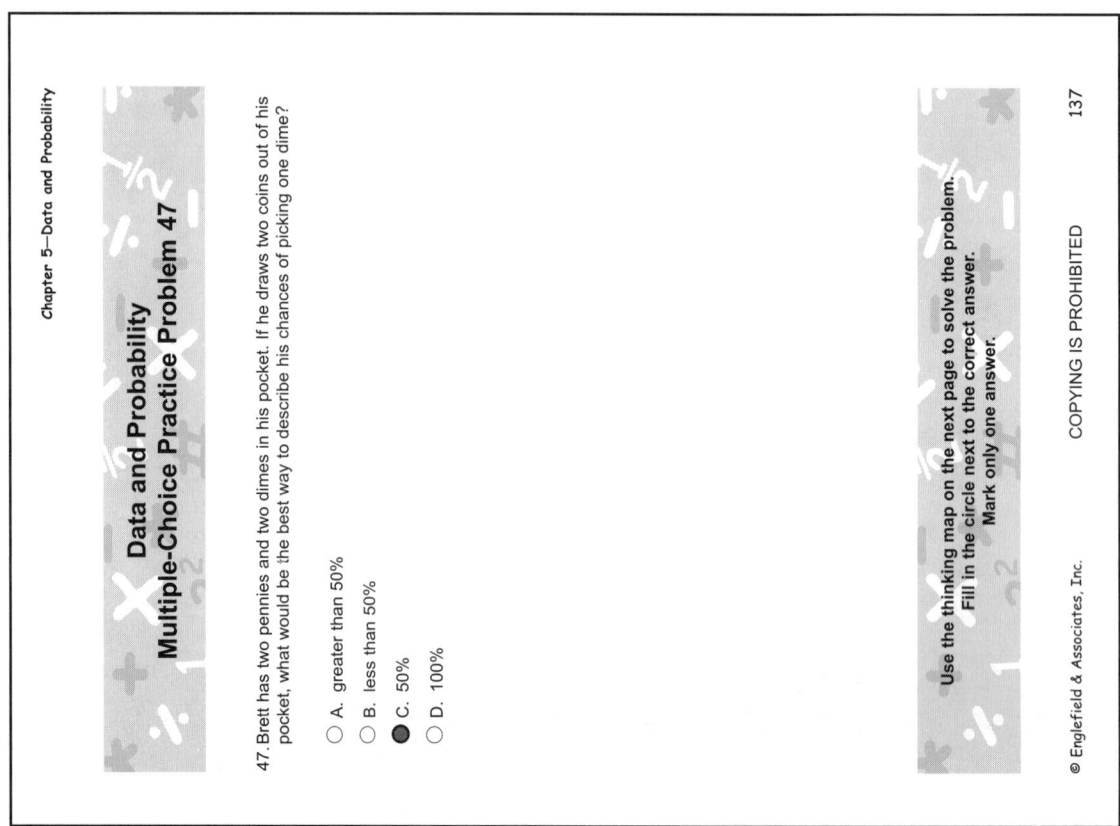

Data and Probability
Multiple-Choice Practice Problem 47

47. Brett has two pennies and two dimes in his pocket. If he draws two coins out of his pocket, what would be the best way to describe his chances of picking one dime?

○ A. greater than 50%
○ B. less than 50%
● C. 50%
○ D. 100%

Use the thinking map on the next page to solve the problem.
Fill in the circle next to the correct answer.
Mark only one answer.

Model Problem 14: Data and Probability Short-Answer

1. Prior Knowledge

Determine the student's prior knowledge about the strand, content of the problem, and/or mathematical process of the problem.

These questions can be used to help determine students' knowledge:

- What are different ways to represent data?
- When have you used data? A graph?

Additional prompts could be provided by using the following topics:

- classroom votes to make a decision
- examining textbooks and newspapers for graphs
- surveying classmates and putting the data in graphs

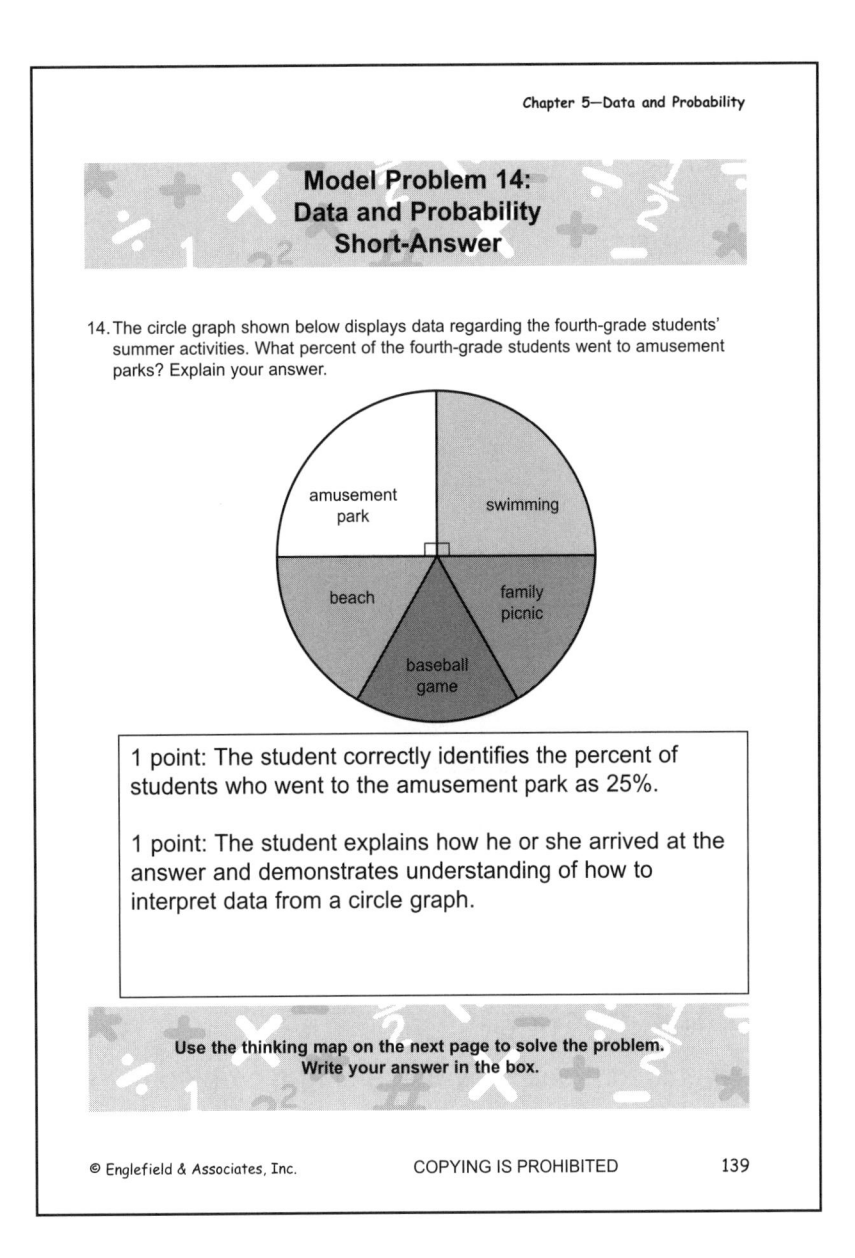

Model Problem 14: Data and Probability Short-Answer

14. The circle graph shown below displays data regarding the fourth-grade students' summer activities. What percent of the fourth-grade students went to amusement parks? Explain your answer.

[Circle graph with sections labeled: amusement park, swimming, beach, family picnic, baseball game]

1 point: The student correctly identifies the percent of students who went to the amusement park as 25%.

1 point: The student explains how he or she arrived at the answer and demonstrates understanding of how to interpret data from a circle graph.

Use the thinking map on the next page to solve the problem. Write your answer in the box.

Chapter 5—Data and Probability

2. Model

Model the use of the mathematical thinking map through class discussion with the thinking process modeled aloud as the thinking map is completed.

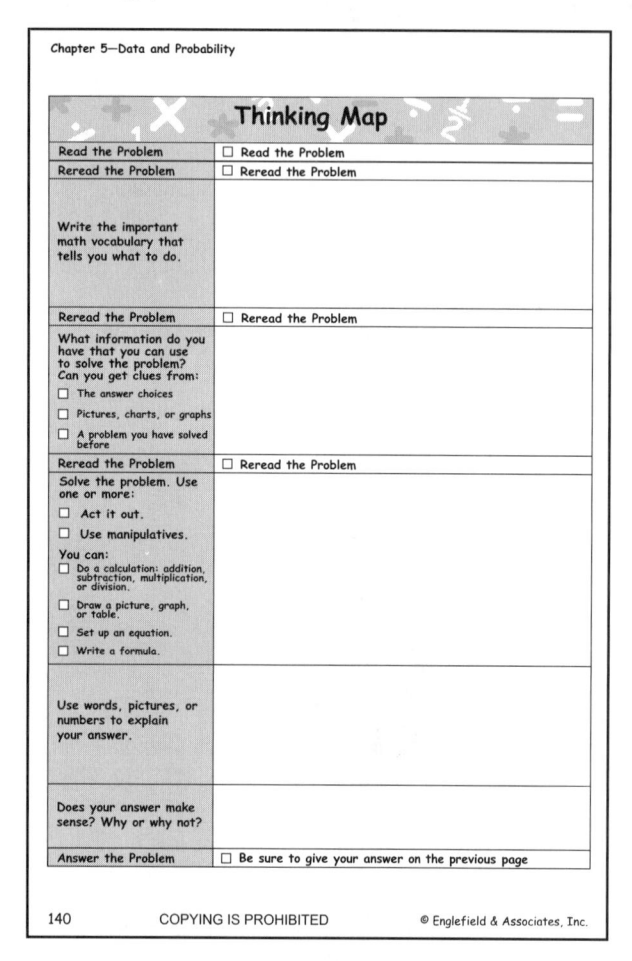

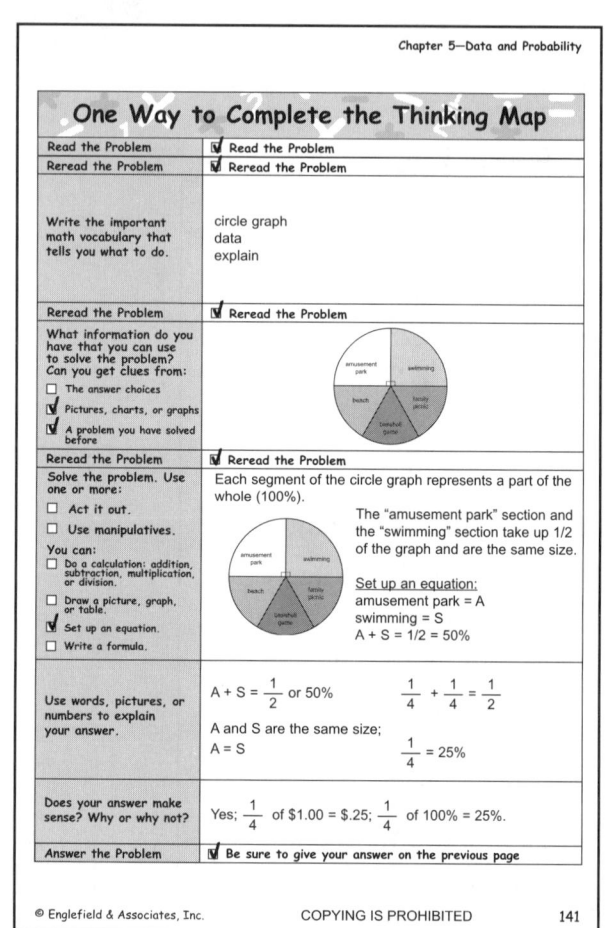

3. Guided Practice

Provide guided practice by presenting a problem for students to complete that matches the strand, content, or process you want to teach.

4. Independent Practice

Provide independent practice so students are provided with multiple-choice, short-answer, or extended-response problems.

Thinking Map

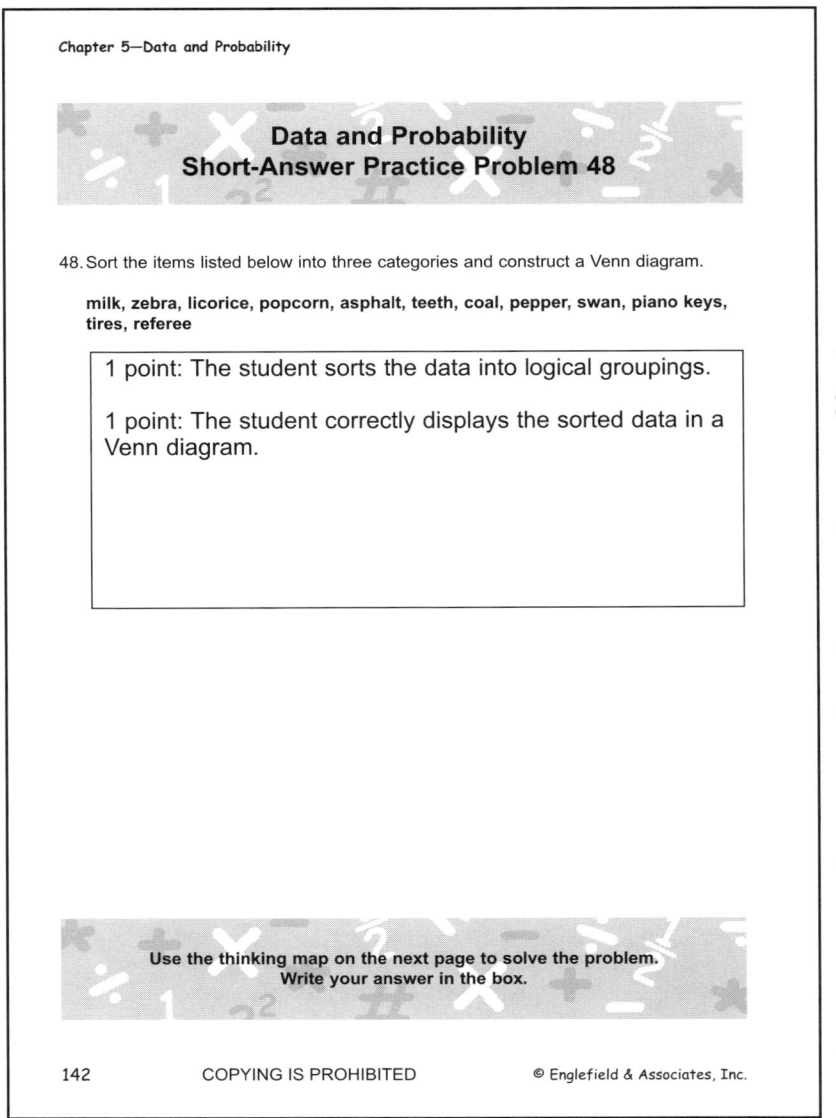

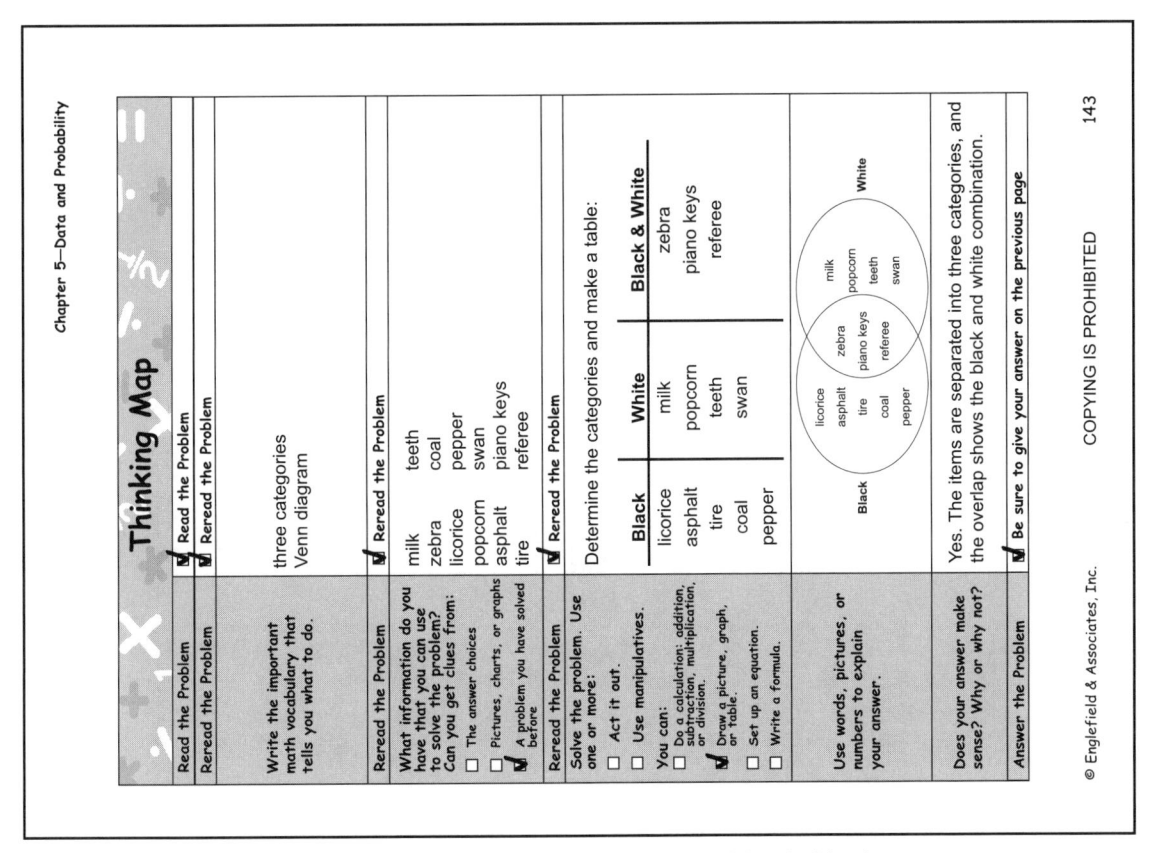

Read the Problem	✓ Read the Problem
Reread the Problem	✓ Reread the Problem
Write the important math vocabulary that tells you what to do.	three categories Venn diagram
Reread the Problem	✓ Reread the Problem
What information do you have that you can use to solve the problem? Can you get clues from: ☐ The answer choices ☑ Pictures, charts, or graphs ☑ A problem you have solved before	milk teeth zebra coal licorice pepper popcorn swan asphalt piano keys tire referee
Reread the Problem	✓ Reread the Problem
Solve the problem. Use one or more: ☐ Act it out. ☐ Use manipulatives. You can: ☐ Do a calculation: addition, subtraction, multiplication, or division. ☑ Draw a picture, graph, or table. ☐ Set up an equation. ☐ Write a formula.	Determine the categories and make a table: | Black	White	Black & White
---	---	---
licorice	milk	zebra
asphalt	popcorn	piano keys
tire	teeth	referee
coal	swan	
pepper			 (Venn diagram showing Black: licorice, asphalt, tire, coal, pepper; overlap: zebra, piano keys, referee; White: milk, popcorn, teeth, swan)
Use words, pictures, or numbers to explain your answer.	Yes. The items are separated into three categories, and the overlap shows the black and white combination.
Does your answer make sense? Why or why not?	✓ Be sure to give your answer on the previous page
Answer the Problem	

In the Student Workbook, the Thinking Map is blank.
In the Parent/Teacher Edition, suggested ways for completing the Thinking Maps are presented.

Data and Probability
Short-Answer Practice Problem 48

48. Sort the items listed below into three categories and construct a Venn diagram.

milk, zebra, licorice, popcorn, asphalt, teeth, coal, pepper, swan, piano keys, tires, referee

1 point: The student sorts the data into logical groupings.

1 point: The student correctly displays the sorted data in a Venn diagram.

Use the thinking map on the next page to solve the problem.
Write your answer in the box.

Chapter 5—Data and Probability

Data and Probability
Short-Answer Practice Problem 49

49. There are five people in a room. Each person shakes hands with every other person. How many handshakes occur? Explain your answer using words or a diagram.

1 point: The student organizes the data in a diagram that shows each person shakes hands with four other people.

1 point: The student correctly computes that there will be 10 total handshakes.

Use the thinking map on the next page to solve the problem. Write your answer in the box.

In the Student Workbook, the Thinking Map is blank.
In the Parent/Teacher Edition, suggested ways for completing the Thinking Maps are presented.

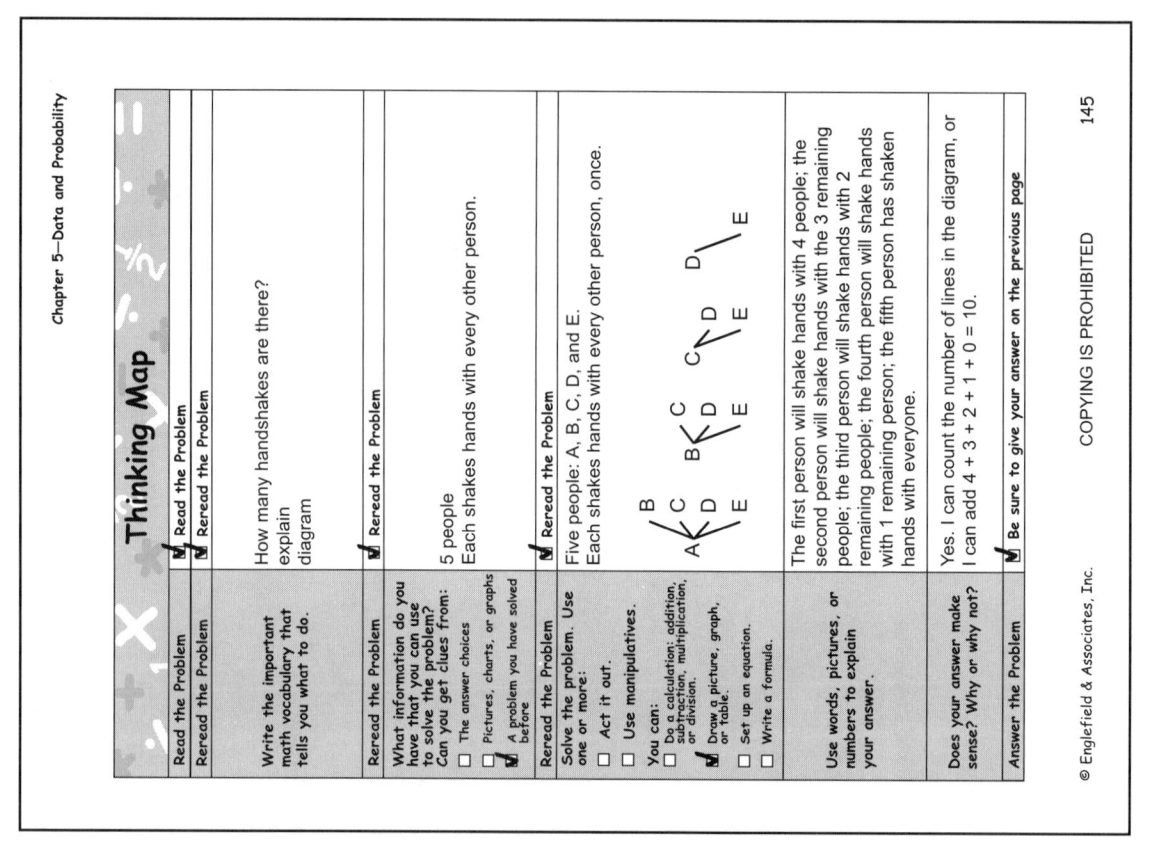

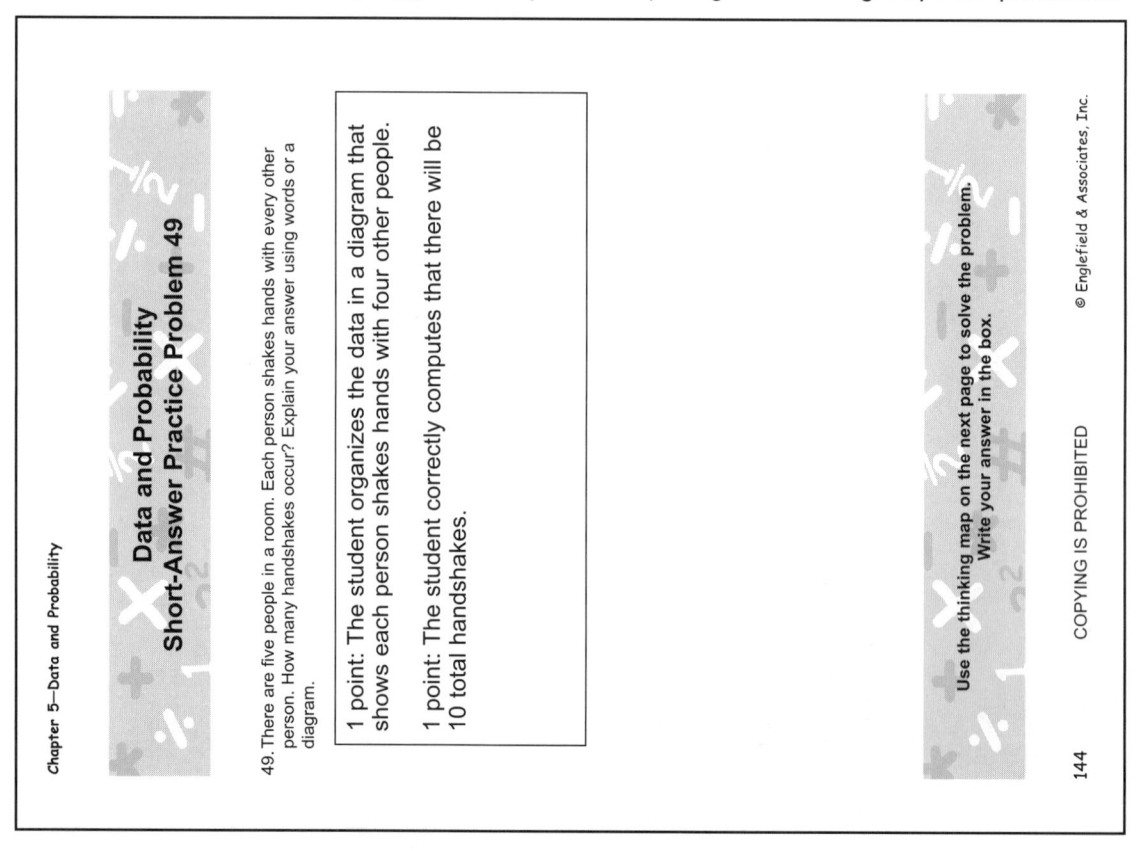

Thinking Map

Read the Problem ✓ Read the Problem
✓ Reread the Problem

Reread the Problem
Write the important math vocabulary that tells you what to do.

How many handshakes are there? explain diagram

Reread the Problem
What information do you have that you can use to solve the problem? Can you get clues from:
☐ The answer choices
☐ Pictures, charts, or graphs
✓ A problem you have solved before

5 people
Each shakes hands with every other person.

Reread the Problem
Solve the problem. Use one or more:
☐ Act it out.
☐ Use manipulatives.
You can:
☐ Do a calculation: addition, subtraction, multiplication, or division.
✓ Draw a picture, graph, or table.
☐ Set up an equation.
☐ Write a formula.

✓ Reread the Problem
Five people: A, B, C, D, and E.
Each shakes hands with every other person, once.

B
C
A — D
E

C
B — D
E

C — D
E

D — E

The first person will shake hands with 4 people; the second person will shake hands with the 3 remaining people; the third person will shake hands with 2 remaining people; the fourth person will shake hands with 1 remaining person; the fifth person has shaken hands with everyone.

Use words, pictures, or numbers to explain your answer.

Does your answer make sense? Why or why not?
Yes. I can count the number of lines in the diagram, or I can add 4 + 3 + 2 + 1 + 0 = 10.

Answer the Problem ✓ Be sure to give your answer on the previous page

Model Problem 15: Data and Probability Extended-Response

1. Prior Knowledge

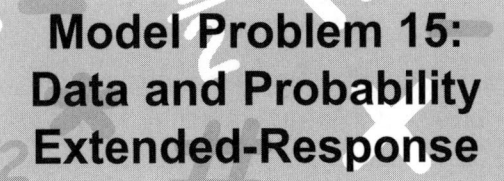

Determine the student's prior knowledge about the strand, content of the problem, and/or mathematical process of the problem.

These questions can be used to help determine students' knowledge:

- When have you used data to help you make a decision?
- How do you determine what type of graph to use when you display your data?

Additional prompts could be provided by using the following topics:

- grades
- sports scores
- average temperatures or rainfall
- current events
- science experiments

Model Problem 15: Data and Probability Extended-Response

15. Create a line graph that shows how many books Yousif read each week during the summer. What does the graph show about Yousif's reading habits?

Week	1	2	3	4	5
# Read	5	4	4	2	1

1 point: The student selects appropriate intervals for displaying data.

1 point: The student correctly constructs and labels a graph.

1 point: The student correctly displays the data on the graph.

1 point: The student draws a reasonable conclusion about the data, such as, "Yousif read less as the summer progressed."

Use the thinking map on the next page to solve the problem. Write your answer in the box.

Chapter 5—Data and Probability

2. Model

Model the use of the mathematical thinking map through class discussion with the thinking process modeled aloud as the thinking map is completed.

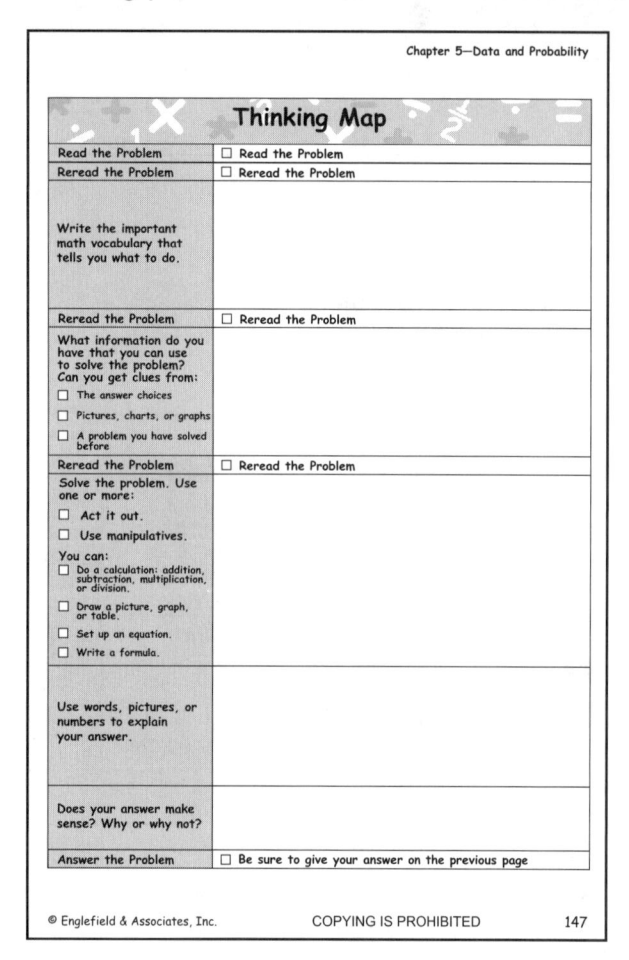

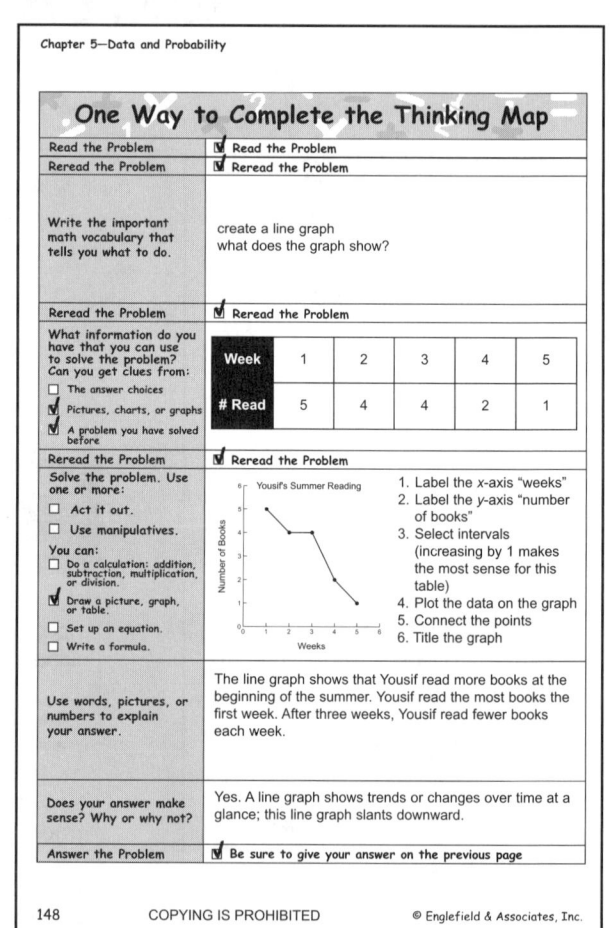

3. Guided Practice

Provide guided practice by presenting a problem for students to complete that matches the strand, content, or process you want to teach.

4. Independent Practice

Provide independent practice so students are provided with multiple-choice, short-answer, or extended-response problems.

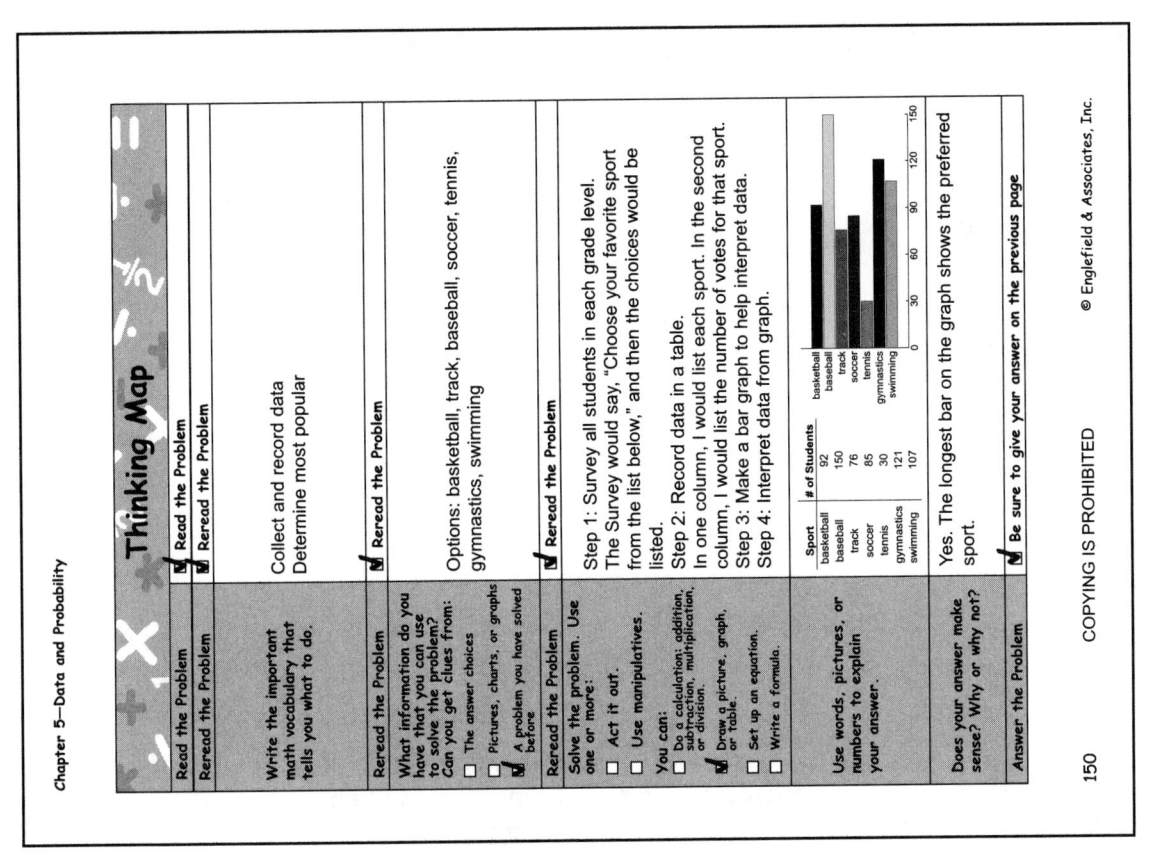

In the Student Workbook, the Thinking Map is blank.
In the Parent/Teacher Edition, suggested ways for completing the Thinking Maps are presented.

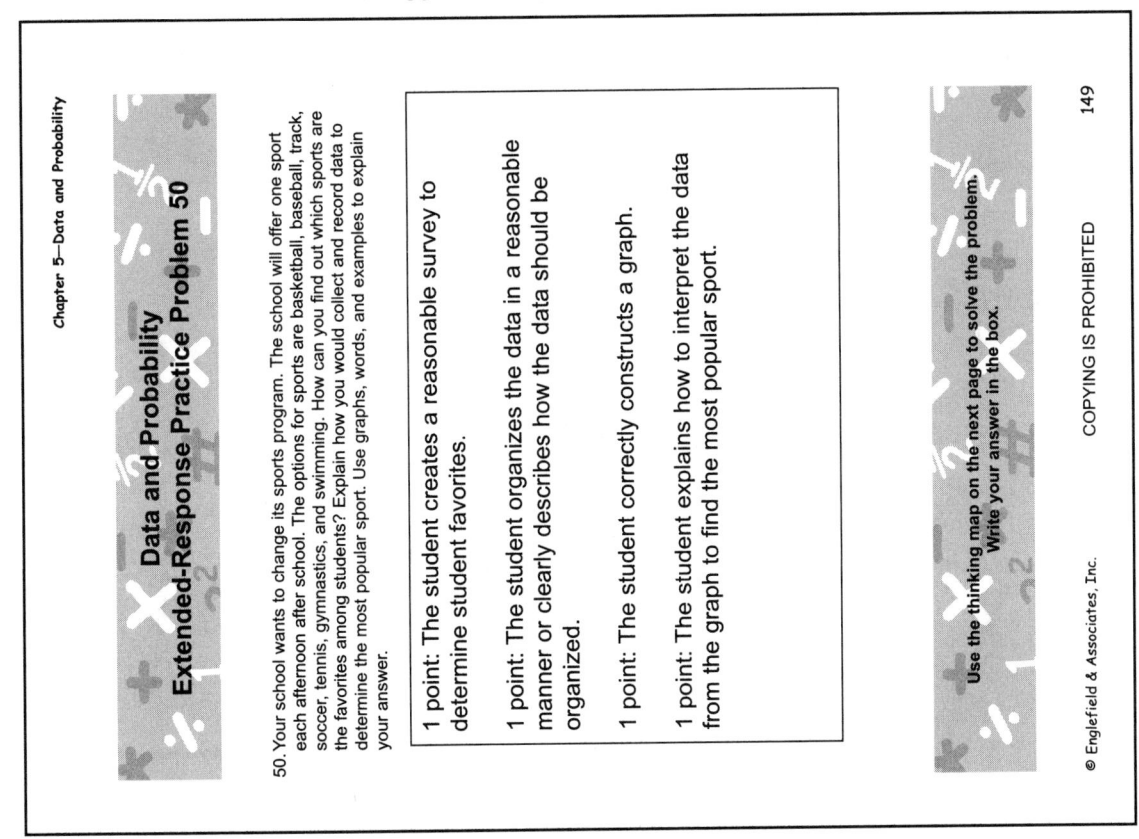

Chapter 5—Data and Probability

Thinking Map

☑ Read the Problem	
☑ Reread the Problem	
Write the important math vocabulary that tells you what to do.	Collect and record data Determine most popular
☑ Reread the Problem	
What information do you have that you can use to solve the problem? Can you get clues from: ☐ The answer choices ☐ Pictures, charts, or graphs ☑ A problem you have solved before	Options: basketball, track, baseball, soccer, tennis, gymnastics, swimming
☑ Reread the Problem	
Solve the problem. Use one or more: ☐ Act it out. ☐ Use manipulatives. You can: ☐ Do a calculation: addition, subtraction, multiplication, or division. ☑ Draw a picture, graph, or table. ☐ Set up an equation. ☐ Write a formula.	Step 1: Survey all students in each grade level. The Survey would say, "Choose your favorite sport from the list below," and then the choices would be listed. Step 2: Record data in a table. In one column, I would list each sport. In the second column, I would list the number of votes for that sport. Step 3: Make a bar graph to help interpret data. Step 4: Interpret data from graph. | Sport | # of Students | |---|---| | basketball | 92 | | baseball | 150 | | track | 76 | | soccer | 85 | | tennis | 30 | | gymnastics | 121 | | swimming | 107 |
Use words, pictures, or numbers to explain your answer.	
Does your answer make sense? Why or why not?	Yes. The longest bar on the graph shows the preferred sport.
☑ Answer the Problem	

150   COPYING IS PROHIBITED   © Englefield & Associates, Inc.

Chapter 5—Data and Probability

Data and Probability
Extended-Response Practice Problem 50

50. Your school wants to change its sports program. The school will offer one sport each afternoon after school. The options for sports are basketball, baseball, track, soccer, tennis, gymnastics, and swimming. How can you find out which sports are the favorites among students? Explain how you would collect and record data to determine the most popular sport. Use graphs, words, and examples to explain your answer.

1 point: The student creates a reasonable survey to determine student favorites.

1 point: The student organizes the data in a reasonable manner or clearly describes how the data should be organized.

1 point: The student correctly constructs a graph.

1 point: The student explains how to interpret the data from the graph to find the most popular sport.

Use the thinking map on the next page to solve the problem.
Write your answer in the box.

Chapter 5—Data and Probability

Literature Connections for Data and Probability

Who's Got Spots?
Linda Williams Aber
© 2000

How Many Feet in the Bed
Diane Johnston Hamm
© 1994

More M&M's Math
Barbara Barbieri McGrath and Roger Glass
© 1998

Do You Wanna Bet?: Your Chance to Find Out About Probability
Jean Cushman
© 1991

The Number Devil: A Mathematical Adventure
Hans Magnus Enzensberger
© 2000

Probably Pistachio: Probability
Stuart J. Murphy
© 2001

In All Probability
Celia Cuomo
© 1998

Guess Who My Favorite Person Is
Byrd Baylor
© 1977

Mysteries & Marvels of Nature
Liz Dalby
© 2003

Ten for Dinner
Jo Ellen Bogart
© 1989

The Great Turkey Walk
Kathleen Karr
© 2000

Rechenka's Eggs
Patricia Polacco
© 1988

Imaginary Numbers: An Anthology of Marvelous Mathematical Stories, Diversions, Poems, and Musings
William Frucht (edited by)
© 1999

Career Ideas for Kids Who Like Math
Diane Lindsey Reeves
© 2000

Jumanji
Chris Van Allsburg
© 1981

Chapter 6

Manipulatives

The manipulatives included on the following pages are tools to help students solve problems. The blackline masters of the manipulatives are provided with the following considerations:

1. Students are provided with tools to help them make abstract mathematical concepts concrete. As students use manipulatives, they can make new discoveries that may not have been considered previously.
2. The manipulatives that are included will help students to connect with the concepts in the problems provided as well as other problems that they may encounter in their regular mathematics program and through standardized tests.
3. Manipulatives provide hands-on opportunities that are both engaging and motivational for students.
4. Manipulatives help students move toward higher-level concepts and problem-solving strategies.

The manipulatives are printed on blackline masters so that teachers, students, and parents can reuse these tools to develop problem-solving skills.

Uses for the Manipulatives

The completed thinking maps included in the Parent/Teacher Edition describe situations in which various manipulatives would help students as they solve the problems included in *Math on Target*.

Two-Dimensional Figures

The two-dimensional figures on page 125 are included to be used as a reference when a problem mentions a particular figure. The shapes can be cut out and used by students to visualize or manipulate to problem solve, or the page can be used as a reference for students.

One-Inch and Half-Inch Grid Paper

Students can use the grids on pages 127 and 129 to construct 3-dimensional figures or to draw areas that require specific dimensions or perspectives. The teacher can select either the one-inch or half-inch grid depending on the problem and the developmental level of the student.

© Englefield & Associates, Inc.

Chapter 6—Manipulatives

Spinners

The spinners are included on page 133 so that parents and teachers can use them to discuss and teach probability. The spinners can be used to build background knowledge as students see what happens when conditions change; for example, in Figure 2 (shown below and also on page 133), the likelihood of landing on the number 3 is 2 out of 4, or 1/2.

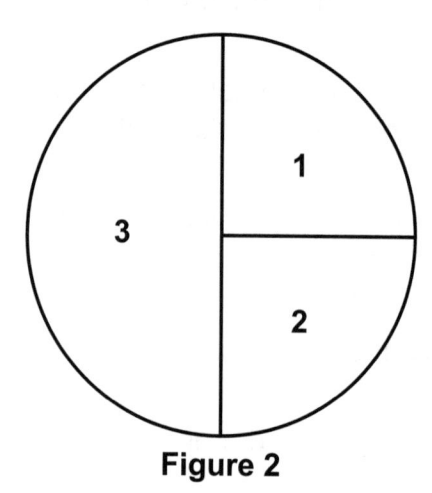

Figure 2

Figures 1, 3, and 4 (page 133) demonstrate equal chances of landing on any of the numbers or colors (if students color them in). These spinners can also be used to play games or experiment with probability as students record the number of times the spinner lands on each of the numbers or colors.

Measurement Tools

The protractor on page 131 is included so that students can measure various angles. The inch and centimeter rulers (also on page 131) are included so that students can use them as a reference for solving problems that involve measurements.

Volume of Cubes

The cube models on page 135 are included so that students have a visual reference for volume to see how volume translates into figures.

Clock (Face with Hands)

The clock model on page 137 is included to allow students to solve problems of change over time.

© Englefield & Associates, Inc.

2-Dimensional Figures

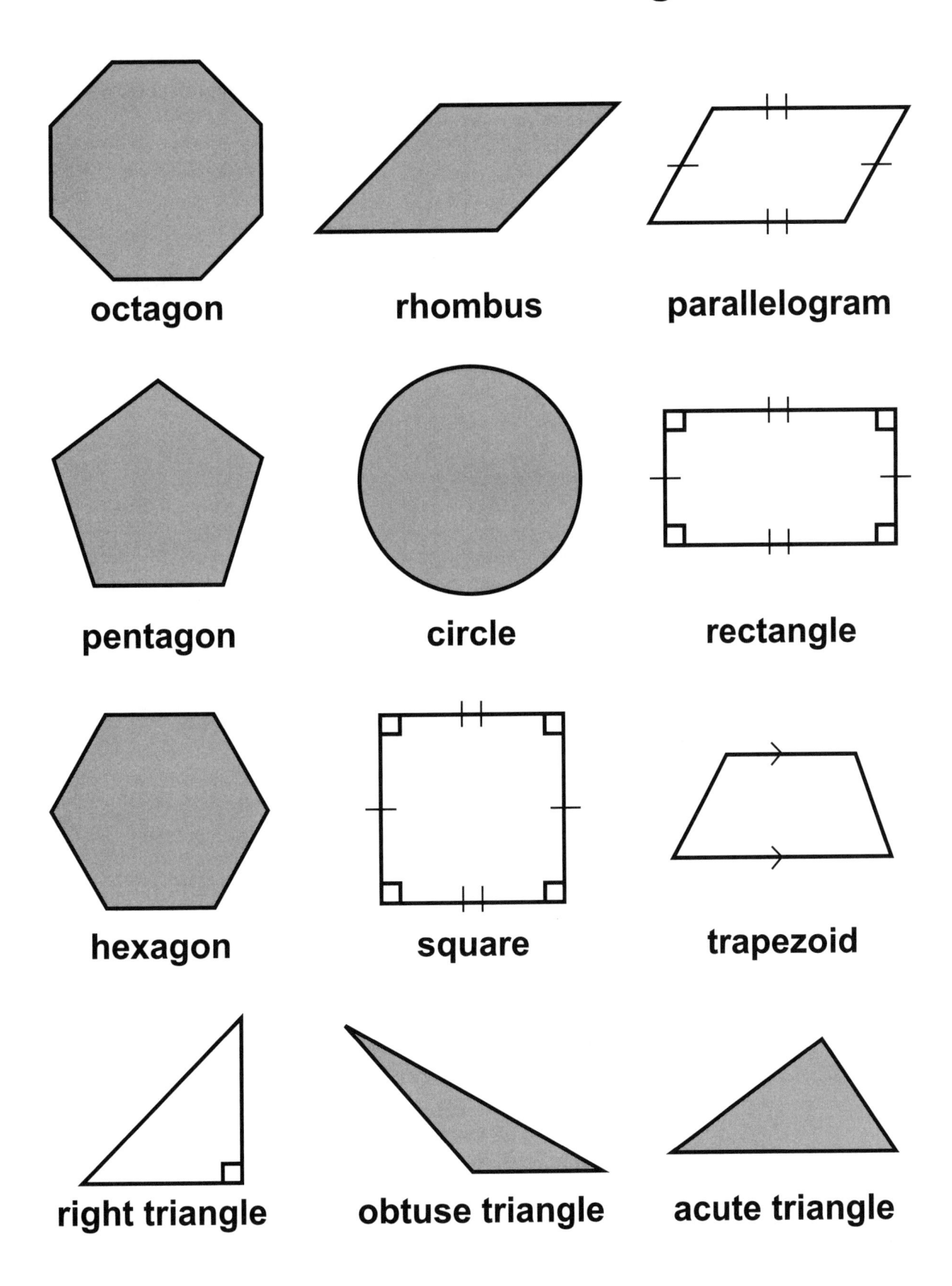

octagon

rhombus

parallelogram

pentagon

circle

rectangle

hexagon

square

trapezoid

right triangle

obtuse triangle

acute triangle

One-Inch Grid

Half-Inch Grid

Chapter 6—Manipulatives

Measurement Tools

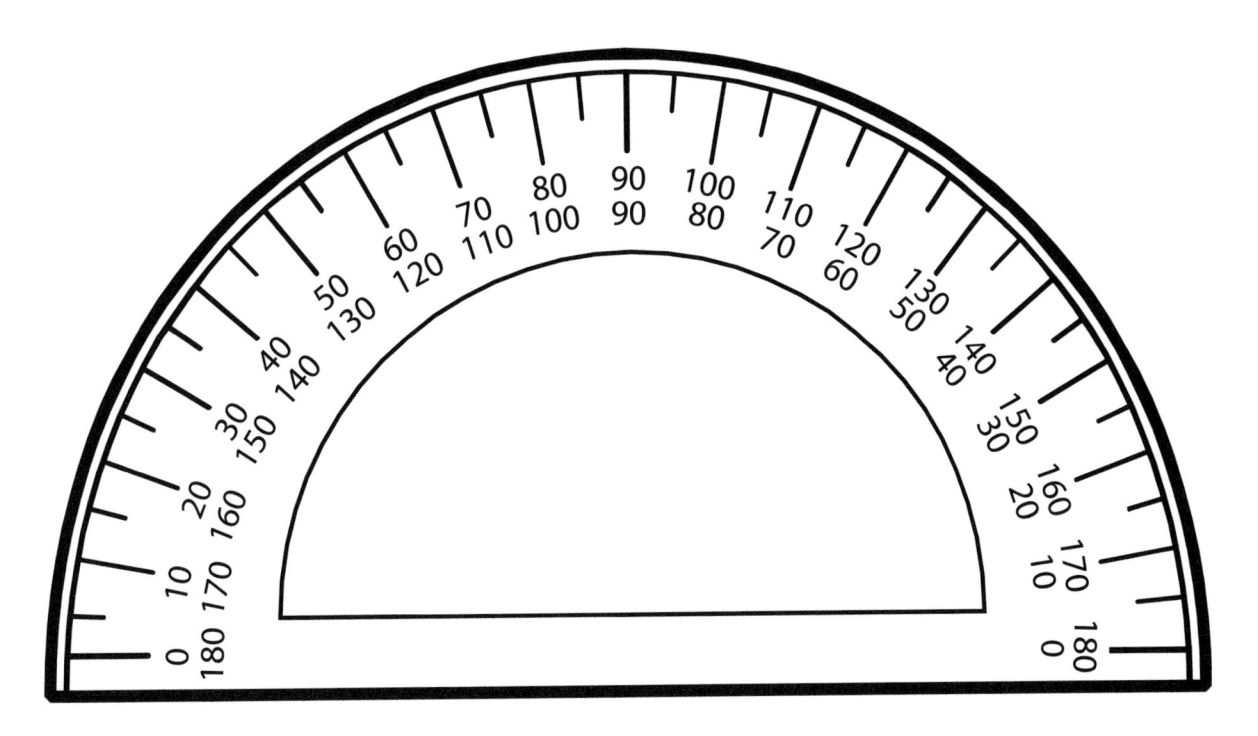

Protractor

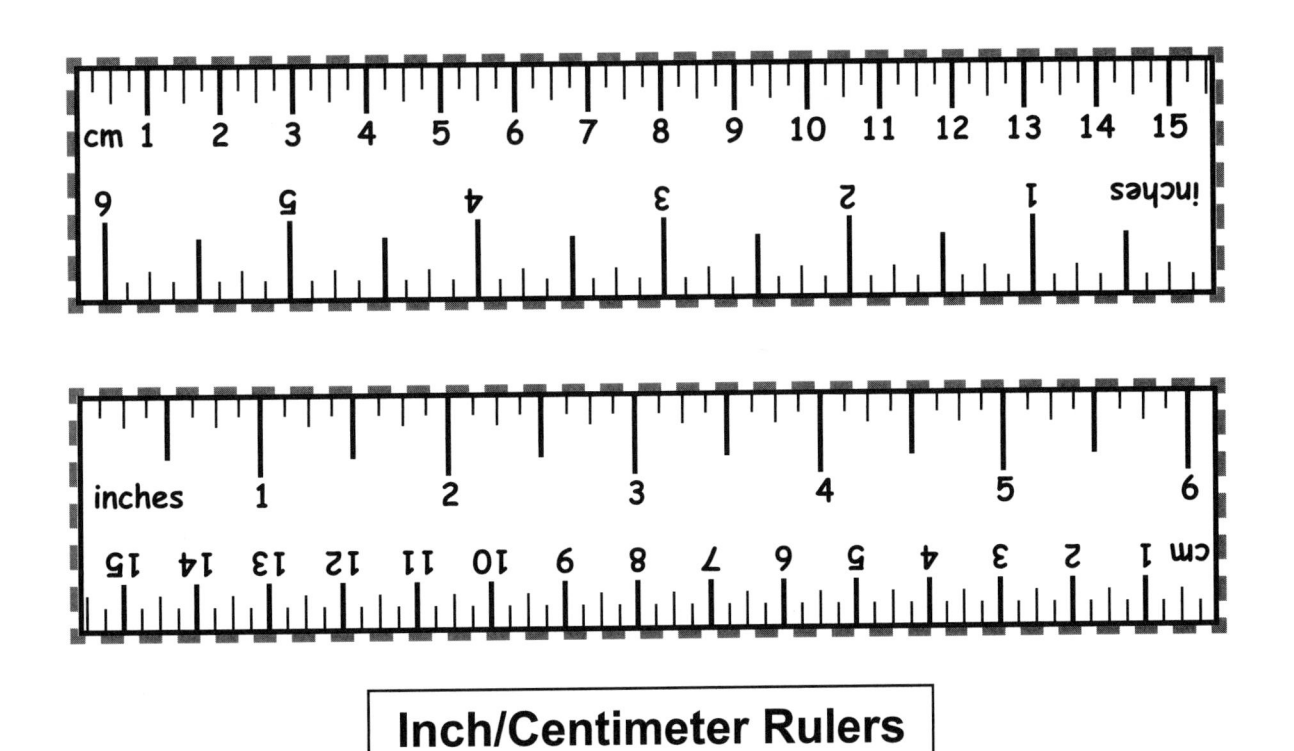

Inch/Centimeter Rulers

© Englefield & Associates, Inc.

Spinners for Probability

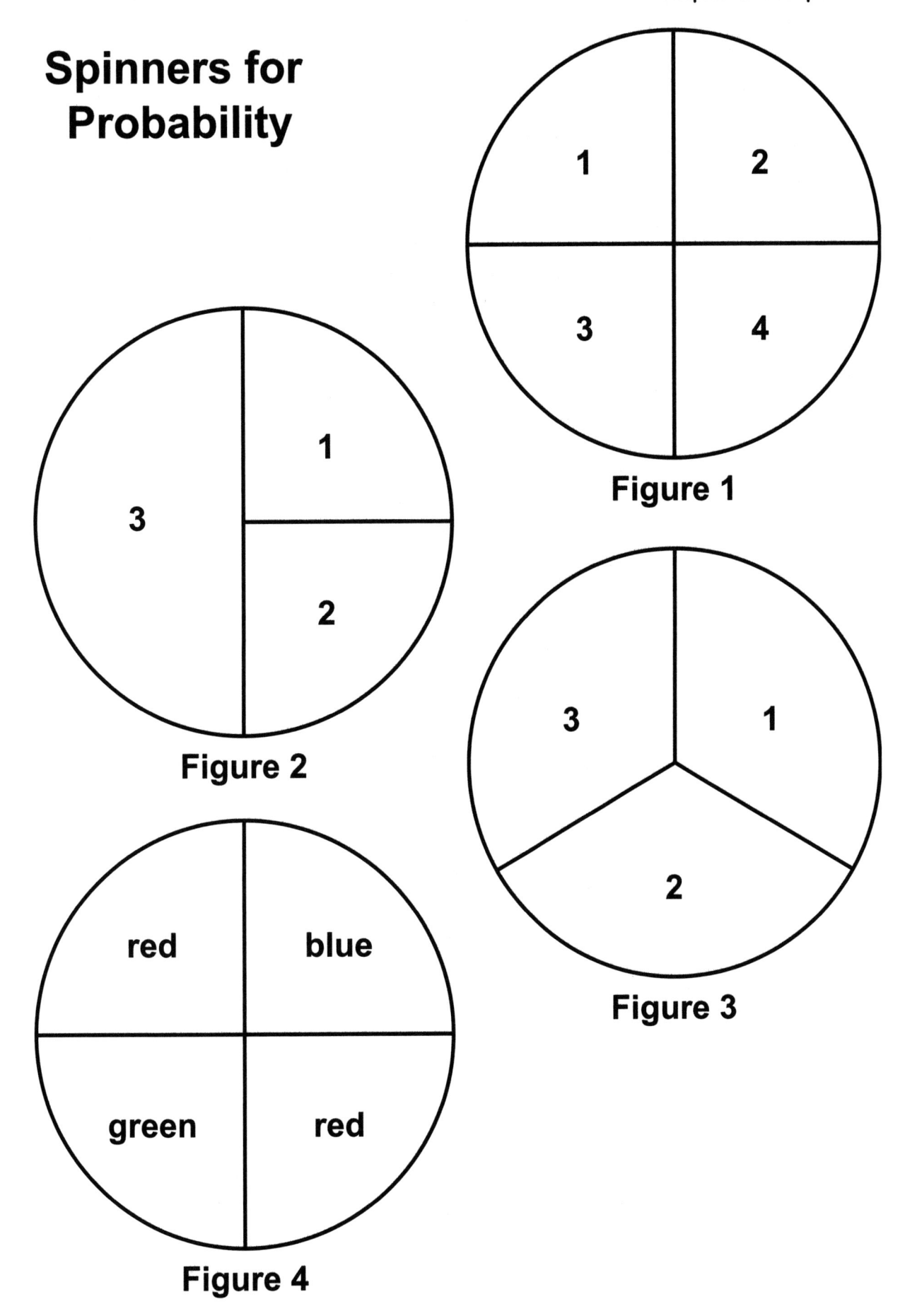

Figure 1 (four equal quadrants): 1, 2, 3, 4

Figure 2 (half labeled 3; other half split into 1 and 2)

Figure 3 (three equal sections): 3, 1, 2

Figure 4 (four equal quadrants): red, blue, green, red

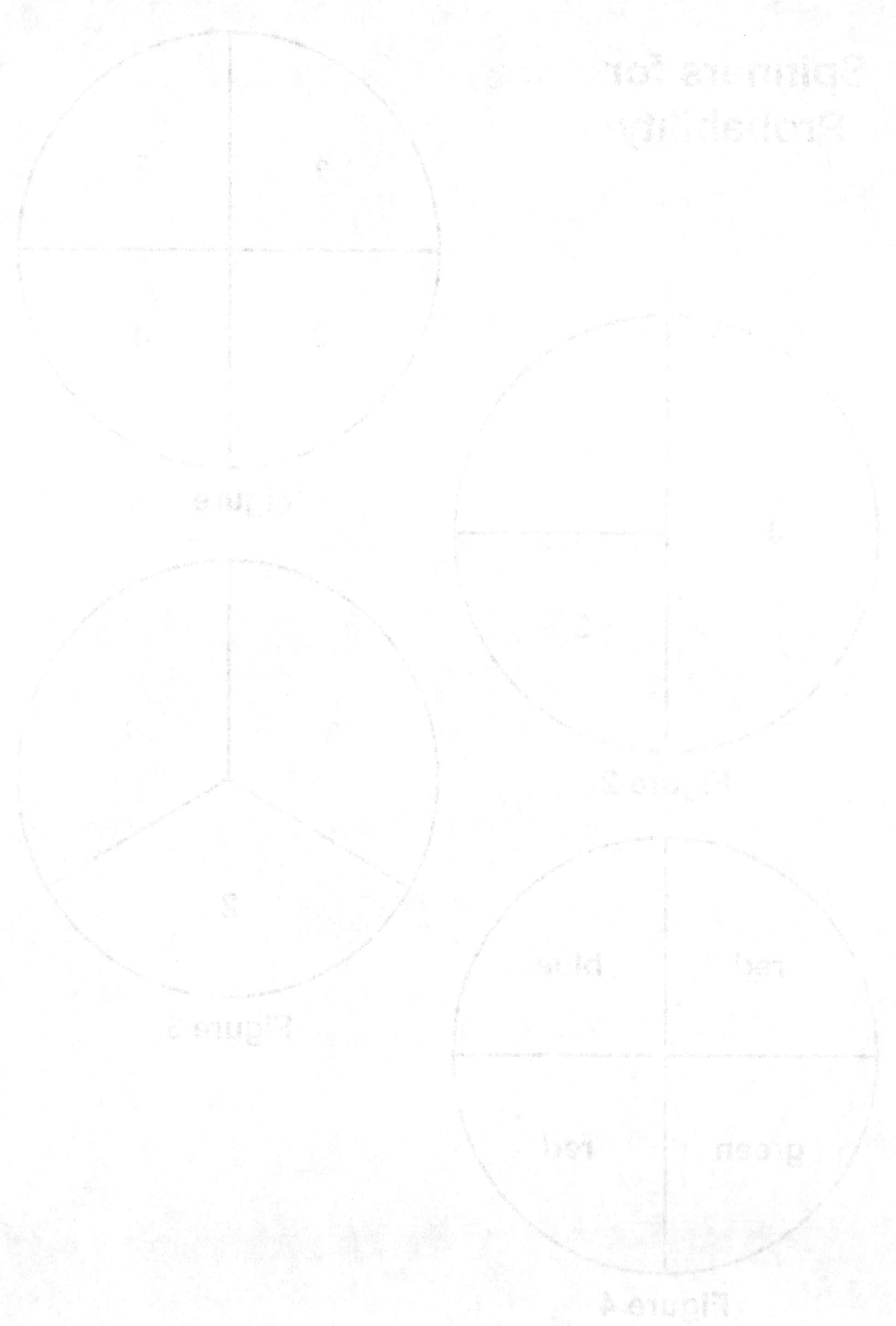

Volume of Cubes

1 cube

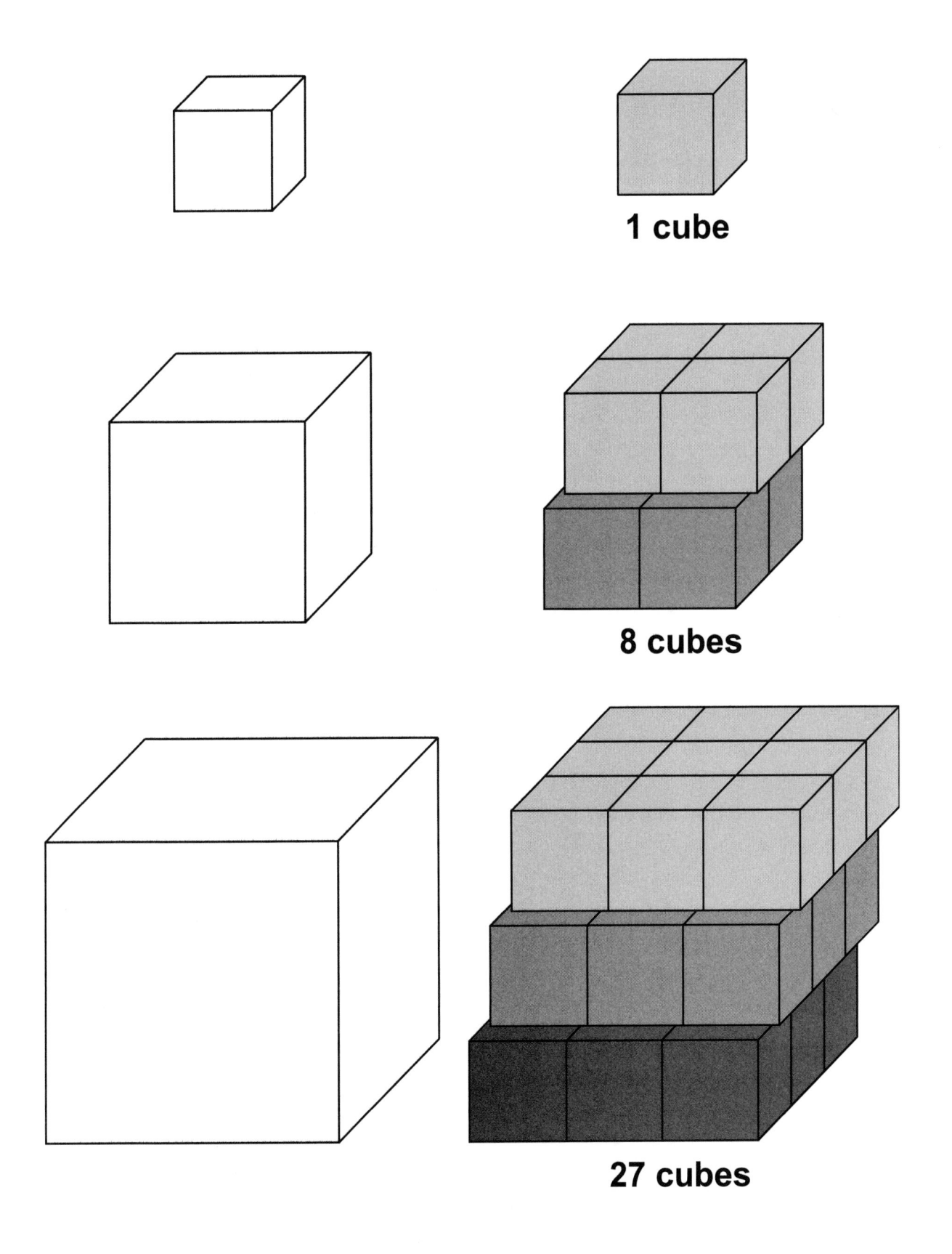

8 cubes

27 cubes

Clock

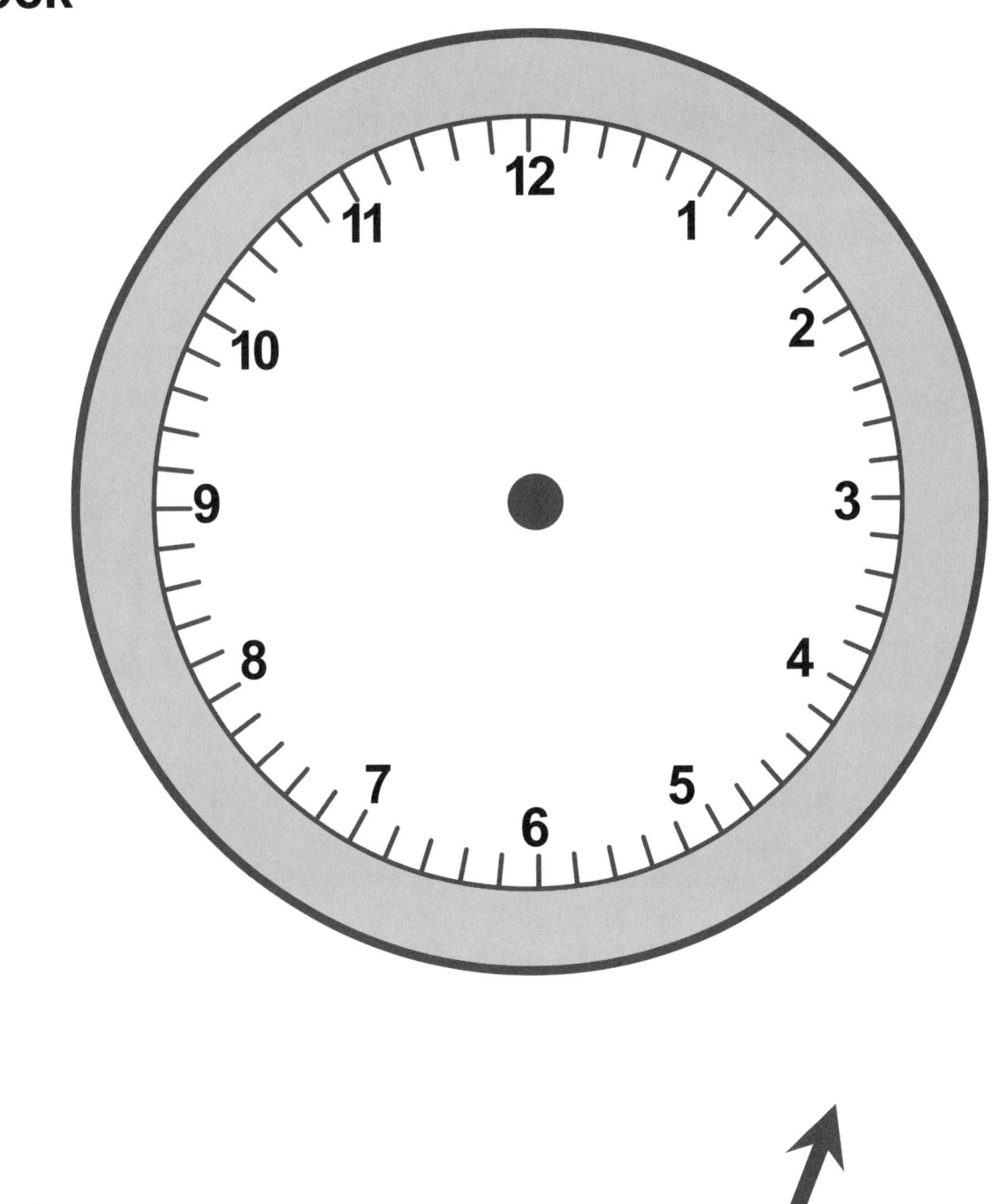

© Englefield & Associates, Inc.

Teacher Notes

Teacher Notes

Teacher Notes

Teacher Notes

Teacher Notes

Subject-Specific Skill Development
Workbooks Increase Testing Skills

Write on Target for grades 1/2, 3, 4, 5, and 6

Includes Graphic Organizers

Read on Target for grades 1/2, 3, 4, 5, and 6

Includes Reading Maps

Math on Target for grades 3, 4, and 5

Includes Thinking Maps

For more information, call our toll-free number: 1.877.PASSING (727.7464)
or visit our website: www.showwhatyouknowpublishing.com